Annals of Mathematics Studies

Number 85

ENTIRE HOLOMORPHIC MAPPINGS IN ONE AND SEVERAL COMPLEX VARIABLES

BY

PHILLIP A. GRIFFITHS

Hermann Weyl Lectures
The Institute for Advanced Study

PRINCETON UNIVERSITY PRESS
AND
UNIVERSITY OF TOKYO PRESS

PRINCETON, NEW JERSEY
1976

Published in Japan exclusively by
University of Tokyo Press;
In other parts of the world by
Princeton University Press

Printed in the United States of America
by Princeton University Press, Princeton, New Jersey

Library of Congress Cataloging in Publication data will
be found on the last printed page of this book

HERMANN WEYL LECTURES

The Hermann Weyl lectures are organized and sponsored by the School of Mathematics of the Institute for Advanced Study. Their aim is to provide broad surveys of various topics in mathematics, accessible to nonspecialists, to be eventually published in the Annals of Mathematics Studies.

The present monograph is the second in this series. It is an outgrowth of the fifth set of Hermann Weyl Lectures, which consisted of five lectures given by Professor Phillip Griffiths at the Institute for Advanced Study on October 31, November 1, 7, 8, 11, 1974.

ARMAND BOREL
JOHN W. MILNOR

TABLE OF CONTENTS

INDEX OF NOTATIONS

C^n is complex Euclidean space with coordinates $z = (z_1, \cdots, z_n)$;

$$(z,w) = \sum_{i=1}^n z_i \bar{w}_i \quad \text{and} \quad \|z\|^2 = (z,z);$$

P^n is complex projective space with homogeneous coordinates
$$Z = [z_0, \cdots, z_n] \; ;$$

$P^1 = C \cup \{\infty\}$ is the Riemannian sphere;

$B[r] = \{z \in C^n : \|z\| < r\}$ is the ball of radius r in C^n;

$S[r] = S \cap B[r]$ for any set $S \subset C^n$;

$\Delta(r) = \{z \in C : |z| < r\}$ is the disc of radius r in C;

$\Delta = \Delta(1)$ is the unit disc;

$\Delta^* = \{\zeta \in C : 0 < |\zeta| < 1\}$ is the punctured disc;

$\Delta^*(R) = \{z \in C^n : |z_i| < R_i \text{ and } R = (R_1, \cdots, R_n)\}$ is a polycylinder in C^n;

$\Delta^*_{k,n} = (\Delta^*)^k \times (\Delta)^{n-k}$ is a punctured polycylinder;

$\Phi, \Psi, \cdots$ denote volume forms;

$\omega, \phi, \psi, \eta, \cdots$ denote $(1,1)$ forms;

$d^c = \dfrac{\sqrt{-1}}{4\pi} (\bar{\partial} - \partial)$;

On C with $z = re^{i\theta}$,

$$d^c = \frac{1}{4\pi} r \frac{\partial}{\partial r} \otimes d\theta - \frac{1}{4\pi} \frac{1}{r} \frac{\partial}{\partial \theta} \otimes dr \; ;$$

$\phi = dd^c \|z\|^2 = \dfrac{\sqrt{-1}}{2\pi} \left(\sum_{i=1}^n dz_i \wedge d\bar{z}_c \right)$ is the standard Kähler form on C^n;

$\omega = dd^c \log \|z\|^2$ is the pull-back to $C^n - \{0\}$ of the Fubini-Study Kähler metric on P^{n-1};

A holomorphic line bundle is denoted by $L \to M$;

$H \to P^n$ is the hyperplane line bundle;

D is a divisor and [D] the corresponding line bundle;

$c_1(L)$ is the Chern form (curvature form) of a Hermitian line bundle;

$\mathcal{O}(M,L)$ is the space of holomorphic sections of $L \to M$;

$|L|$ is the projective space of divisors of sections $s \in \mathcal{O}(M,L)$;

$H^2_{DR}(M,R)$ is the 2nd deRham cohomology of M;

A real (1,1) form ψ on M is positive in case locally

$$\psi = \frac{\sqrt{-1}}{2} \sum \psi_{ij} \, dz_i \wedge d\bar{z}_j$$

where (ψ_{ij}) is a positive definite Hermitian matrix;

$[\psi]$ denotes the class in $H^2_{DR}(M,R)$ of a closed form ψ on M;

A class χ in $H^2_{DR}(M,R)$ is positive in case $\chi = [\psi]$ for some positive form ψ;

K_M is the canonical line bundle of M;

L^* is the dual line bundle to $L \to M$;

$\mathcal{O}(C^n)$ are the entire holomorphic functions on C^n.

Entire Holomorphic Mappings
in One and Several
Complex Variables

ENTIRE HOLOMORPHIC MAPPINGS IN ONE AND SEVERAL COMPLEX VARIABLES

Phillip A. Griffiths[*]

INTRODUCTION

(a) *Some general remarks*

These talks will be concerned with the value distribution theory of an *entire holomorphic mapping*

$$f : \mathbf{C}^n \to M$$

where M is a compact, complex manifold. The theory began with R. Nevanlinna's quantitative refinement of the Picard theorem concerning a non-constant entire meromorphic function

$$f : \mathbf{C} \to \mathbf{P}^1 .$$

If we let $n_f(a, r)$ be the number of solutions to the equation

$$f(z) = a, \quad |z| < r \text{ and } a \, \epsilon \, \mathbf{P}^1 ,$$

then Picard's theorem says that the sum

$$n_f(a, r) + n_f(b, r) + n_f(c, r)$$

is eventually positive, where a, b, c are three distinct points on $\mathbf{P}^1$. Roughly speaking, Nevanlinna's refinement states that the above sum is eventually larger than the average

$$t_f(r) = \int_{a \, \epsilon \, \mathbf{P}^1} n_f(a, r) \, da$$

[*]This work was partially supported by NSF Grant GP38886.

number of solutions to the equation in question. Since the appearance of
Nevanlinna's book [31], there has been considerable attention to the subject of value distribution theory, first in the classical case of an entire
meromorphic function [25], then in the study of entire holomorphic curves
in projective space ([44] and [45]), and, in recent years, in the general
theory of holomorphic mappings between arbitrary complex manifolds (cf.
[37] for a survey).

We shall concentrate on the *equidimensional case* where $\dim_{\mathbf{C}} M = n$
and where f is *non-degenerate* in the sense that the Jacobian determinant
of f is non-identically zero. Aside from the Ahlfors theory of holomorphic
curves in $\mathbf{P}^n$, it is here that the defect relations of R. Nevanlinna have
been most directly generalized. For example, Picard's theorem becomes
the assertion that no such f can omit a divisor on M which has simple
normal crossings and whose Chern class is larger than that of the anti-
canonical divisor on M. The corresponding defect relation will be proved
in Chapter 3.

Aside from proving this theorem, there are two main purposes of these
lectures. The first is to attempt to integrate more closely the deeper
analytic aspects of the classical one variable theory with the formalism
and algebro-geometric flavor in the several variable case. The second
is to try to isolate the analytic concept of *growth* and differential-
geometric notion of *negative curvature* as being perhaps most basic to the
theory. A glance at the table of contents should make it pretty clear how
our discussion has been centered around these two purposes.

One other aspect of these notes is that we have tried to give the
heuristic reasoning which historically led to the recognition that growth
and curvature were central to theory. Aside from the original proof of
Picard's theorem using the modular function, we have included three
additional proofs, among them the "elementary proof" of Emile Borel in
which the central importance of growth was first clearly exhibited. Similarly,
aside from the negative curvature derivation of the general defect relation,

we have included the two classical proofs due to R. Nevanlinna and Ahlfors, each of which has its own distinct merit.

(b) *General references and background material*

The classic book in the subject is [31] by R. Nevanlinna. A second book of his [32] and the more recent monograph by Hayman [25] contain further discussion of the value distribution theory of an entire meromorphic function. All that is required to read any of these books is standard basic knowledge of complex function theory.

In several variables, we shall use the formalism of complex manifold theory, especially that centered around divisors, line bundles, and their Chern forms and subsequent Chern classes. The basic references here are the books by Chern [11] and Wells [43]. In practice all that we shall really require is fully explained in the introduction (pp. 151-155) of Griffiths-King [22].

From several complex variables, one needs to know a little about the local structure of an analytic hypersurface, especially as regards integration over an analytic hypersurface and Stokes' theorem in this situation. For the former the second chapter of Gunning and Rossi [24] is more than sufficient, and for integration we suggest the notes by Stolzenberg [41].

Aside from the facts that integration of a smooth differential form over a possibly singular analytic variety is possible and that Stokes' formula is valid, the fundamental result we shall use concerning integration on complex varieties is the *Wirtinger theorem:*

Let $ds^2 = \sum_{i,j} h_{ij} dz_i d\bar{z}_j$ *and* V *be respectively a Hermitian metric and k-dimensional analytic variety, both defined in some neighborhood of the closure of an open set* $U \subset C^n$. *Let* $\phi = \dfrac{\sqrt{-1}}{2} \sum_{i,j} h_{ij} dz_i \wedge d\bar{z}_j$ *be the exterior* (1,1) *form associated to the Hermitian metric and* vol(V) *the 2k-dimensional volume of* $V \cap U$ *computed with respect to the Riemannian metric associated to* ds^2. *Then*

$$(0.1) \qquad\qquad \text{vol}\,(V) = \frac{1}{k!} \int\limits_{V \cap U} \phi^k$$

where $\phi^k = \underbrace{\phi \wedge \cdots \wedge \phi}_{k\text{-times}}$.

PROOF. If V^* is the set of non-singular points of V, then by definition $\text{vol}\,(V) = \text{vol}\,(V^*)$ and $\int\limits_{V \cap U} \phi^k = \int\limits_{V^* \cap U} \phi^k$. Thus it suffices to prove (0.1) in case V is a complex submanifold of U, and this obviously reduces to establishing (0.1) in case $V = U$. By applying the Gram-Schmidt process to the differentials $dz_1, \cdots, dz_n$ relative to the Hermitian metric (h_{ij}), we may write

$$ds^2 = \sum_{i=1}^{n} \phi_i \bar{\phi}_i$$

where the $\phi_i = \sum_j a_{ij} dz_j$ give a C^∞ basis for the $(1,0)$ forms over U. Expanding

$$\omega_i = \alpha_i + \sqrt{-1}\,\beta_i$$

in real and imaginary parts, the associated Riemannian metric is

$$ds^2 = \sum_{i=1}^{n} \alpha_i^2 + \beta_i^2 \ .$$

The volume form for this metric is, by definition,

$$d\mu = \alpha_1 \wedge \beta_1 \wedge \cdots \wedge \alpha_n \wedge \beta_n \ .$$

On the other hand,

$$\phi = \frac{\sqrt{-1}}{2}\left(\sum_{i=1}^{n} \phi_i \wedge \bar{\phi}_i \right)$$

so that

$$\phi^n = \left(\frac{\sqrt{-1}}{2}\right)^n n!\ \phi_1 \wedge \bar{\phi}_1 \wedge \cdots \wedge \phi_n \wedge \bar{\phi}_n$$

$$= n!\ \alpha_1 \wedge \beta_1 \wedge \cdots \wedge \alpha_n \wedge \beta_n$$

$$= n!\ d\mu \ . \qquad\qquad\qquad \text{Q.E.D.}$$

The principle that the volume of an analytic variety $V \subset \mathbf{C}^n$ is computed by integration of a differential form defined on all of $\mathbf{C}^n$, as opposed to the computation of arclength, surface area, etc. in the real case, is of fundamental importance.

CHAPTER 1

ORDERS OF GROWTH

(a) *Some heuristic comments*

In its simplest terms, the study of entire holomorphic mappings is concerned with *growth*. For example, when the well-known growth behavior of a polynomial is plugged into the Cauchy integral formula

$$n_p(0, r) = \frac{1}{2\pi \sqrt{-1}} \int\limits_{|z| = r} \frac{p'(z)\, dz}{p(z)}$$

one obtains the fundamental theorem of algebra. Emile Borel realized that growth was also essential to an understanding of *Picard's theorem*, viewed as a transcendental analogue of the fundamental theorem of algebra. Here is the heuristic reasoning behind his proof of that theorem:

An entire holomorphic mapping $f : C \rightarrow P^1$ may be written in homogeneous coordinates as $f(z) = [f_0(z), f_1(z)]$ where f_0 and f_1 are entire holomorphic functions having no common zeroes. We assume that f is nonconstant and omits three points, which may be taken to be $[1, 0]$, $[0, 1]$, and $[1, -1]$. Then

$$\begin{cases} f_0 = e^{h_0}, \ f_1 = e^{h_1} \\ \text{and} \\ e^{h_0}/e^{h_1} + 1 = e^{h_2} \end{cases}$$

where h_0, h_1 and h_2 are entire holomorphic functions. Multiplying the second equation by e^{-h_2} gives a linear relation

8

$$e^{g(z)} + e^{h(z)} = 1$$

between exponentials of entire holomorphic functions. Taking derivatives
yields

$$g'(z) e^{g(z)} + h'(z) e^{h(z)} = 0 ,$$

and these equations may be solved to obtain

$$\begin{cases} e^g = \dfrac{h'}{h' - g'} \\[2ex] e^h = \dfrac{g'}{g' - h'} \end{cases} .$$

In the simplest situation when g and h are polynomials, such a relation
can hold only if $g' \equiv h'$, in which case f is constant. In general, it
seems at least plausible that the left hand side should in some sense grow
exponentially faster than the right hand side. Making this precise necessi-
tates a careful discussion of growth, a discussion which should include
growth of entire meromorphic as well as holomorphic functions, and which
should also relate the growth of a function to that of its derivatives. This
will be the object of the first chapter, with the completion of Borel's argument appearing at the end of Section (d).

Before going on to do this, we want to derive *Jensen's theorem*, which
together with its variants, the First and Second Main Theorems, constitutes
the main tool in the subject. Given an entire holomorphic function $f(z)$,
the number of solutions $n(a, r)$ to the equation

$$f(z) = a$$

in the disc $|z| < r$ is given by the Cauchy formula

$$n(a, r) = \frac{1}{2\pi \sqrt{-1}} \int\limits_{|z|=r} \frac{f'(z)\, dz}{f(z) - a} .$$

The difficulty in using this to estimate $n(a,r)$ is that the integrand is complex while $n(a,r)$ is real, which suggests adding to it the conjugate formula. To facilitate doing this, we write

$$
\begin{aligned}
\frac{f'(z)\,dz}{f(z)-a} &= \partial \log[f(z)-a] \\
&= \partial \log|f(z)-a|^2
\end{aligned}
$$

since $\partial \log(\overline{f(z)-a}) = 0$ by the Cauchy-Riemann equations. It follows that

$$
\begin{aligned}
n(a,r) &= \frac{1}{4\pi\sqrt{-1}} \int_{|z|=r} (\partial - \bar\partial)\log|f(z)-a|^2 \\
&= \int_{|z|=r} d^c \log|f(z)-a|^2
\end{aligned}
$$

where

$$
d^c = \frac{\sqrt{-1}}{4\pi}(\bar\partial - \partial) = \frac{1}{4\pi} r \frac{\partial}{\partial r} \otimes d\theta - \frac{1}{4\pi r}\frac{\partial}{\partial \theta} \otimes dr \; .
$$

Plugging this in gives

$$
n(a,r) = r \frac{d}{dr}\left(\frac{1}{4\pi} \int_{|z|=r} \log|f(z)-a|^2\, d\theta \right) ,
$$

an equation whose form suggests that we integrate with respect to $\frac{dr}{r}$. Setting

$$
N(a,r) = \int_0^r \{n(a,\rho) - n(a,0)\}\frac{d\rho}{\rho} + n(a,0)\log r
$$

we obtain *Jensen's theorem*:

$$
(1.1) \qquad \log|f(0)-a| + N(a,r) = \frac{1}{2\pi}\int_0^{2\pi} \log|f(re^{i\theta})-a|\,d\theta \; ,
$$

valid for an entire holomorphic function $f(z)$ with $f(0) \neq a$. In the general case when $f(z)$ is meromorphic and $f(0) = a$ or ∞, the same proof gives

(1.2) $\qquad N(a,r) = \frac{1}{2\pi} \int_0^{2\pi} \log|f(re^{i\theta}) - a| \, d\theta + N(\infty, r) + C ,$

where

$$C = \lim_{\varepsilon \to 0} \frac{1}{2\pi} \int_0^{2\pi} \log|f(\varepsilon e^{i\theta}) - a| \, d\theta .$$

The formula (1.1) gives an upper bound on $N(a,r)$, and hence on $n(a,\rho)$ for $\rho < r$. It fails, however, to give a lower bound unless we know something about the *mean proximity*

$$\frac{1}{2\pi} \int_0^{2\pi} \log|f(re^{i\theta}) - a| \, d\theta$$

of $f(z) - a$ to infinity on large circles. For example, if $f(z) = e^z$ and $a = 0$, then $f(re^{i\theta})$ tends to ∞ for $-\pi/2 < \theta < \pi/2$ and symmetrically to 0 for $\pi/2 < \theta < 3\pi/2$, and thus the mean proximity is zero. In some sense, understanding integrals of the general form

$$\frac{1}{2\pi} \int_\alpha^\beta \log|f(re^{i\theta})| \, d\theta \qquad (\alpha < \beta)$$

is what the theory is all about. What is not so clear is how the concept of negative curvature appears naturally in this mean proximity problem, and this will be taken up in the second chapter.

(b) *Order of growth of entire analytic sets*

We now turn to a precise analytical description of growth, first for analytic varieties and then for holomorphic mappings.

Suppose that $V \subset \mathbf{C}^n$ is an *entire analytic set* of pure dimension k. Denote by $B[r] = \{z \in \mathbf{C}^n : \|z\| < r\}$ the ball of radius r in $\mathbf{C}^n$ and $V[r] = V \cap B[r]$ that part of V in $B[r]$. (Exception: When $n = 1$, we let $\Delta(r)$ denote the disc of radius r in $\mathbf{C}$.) The Euclidean volume

$$\text{vol}\,(V[r])$$

of $V[r]$ is finite, and we set

$$
\begin{cases}
n(V,r) = C_k \, \mathrm{vol}\,(V[r]) \\[2mm]
\nu(V,r) = \dfrac{n(V,r)}{r^{2k}} \, ,
\end{cases}
$$

where the constant C_k is determined so that $\nu(C^k,r) \equiv 1$.

(1.3) PROPOSITION: *The function $\nu(V,r)$ is increasing in r, and*

$$
\lim_{r \to 0} \nu(V,r) = \nu(V,0)
$$

is the multiplicity of V at the origin.

PROOF. We consider the $(1,1)$ forms

$$
(1.4) \quad
\begin{cases}
\phi = dd^c \|z\|^2 = \dfrac{\sqrt{-1}}{2\pi}\left(\displaystyle\sum_{i=1}^{n} dz_i \wedge d\bar{z}_i \right) = \dfrac{\sqrt{-1}}{2\pi}\,(dz,dz) \\[4mm]
\omega = dd^c \log \|z\|^2 = \dfrac{\sqrt{-1}}{2\pi}\,\dfrac{(z,z)(dz,dz) - (dz,z)(z,dz)}{\|z\|^2} \, .
\end{cases}
$$

The first one ϕ is the standard Kähler form on C^n, and by *Wirtinger's theorem* (0.1) (omitting the factorials)

$$
\mathrm{vol}\,(V[r]) = \int_{V[r]} \phi^k \, .
$$

The second form ω is the pullback to $C^n - \{0\}$ of the standard Kähler form on the P^{n-1} of lines through the origin in C^n. We note that on the tangent space to the sphere $\|z\| = r$

$$
\begin{cases}
\dfrac{1}{r^2}\, d^c \|z\|^2 = d^c \log \|z\|^2 \\[4mm]
\dfrac{1}{r^2}\, dd^c \|z\|^2 = dd^c \log \|z\|^2 & \text{(since } d\|z\|^2 = 0 \text{ there)} \, .
\end{cases}
$$

Now use these remarks together with (1.4) and Stokes' theorem to calculate, for $r < R$,

$$\nu(V,R) - \nu(V,r) = \frac{1}{R^{2k}} \int_{V[R]} \phi^k - \frac{1}{r^{2k}} \int_{V[r]} \phi^k$$

$$= \frac{1}{R^{2k}} \int_{\partial V[R]} d^c \|z\|^2 \wedge \phi^{k-1} - \frac{1}{r^{2k}} \int_{\partial V[r]} d^c \|z\|^2 \wedge \phi^{k-1}$$

$$= \int_{\partial V[R]} d^c \log \|z\|^2 \wedge \omega^{k-1} - \int_{\partial V[r]} d^c \log \|z\|^2 \wedge \omega^{k-1}$$

$$= \int_{V[r,R]} \omega^k$$

where $V[r,R]$ is the intersection of V with the annular ring $r \leq \|z\| \leq R$. This proves that $\nu(V,r)$ is non-decreasing in r. To evaluate $\lim_{r \to 0} \nu(V,r)$, we write for $r < R$

$$\nu(V,r) = \int_{\partial V[r]} d^c \log \|z\|^2 \wedge \omega^{k-1} + \left\{ \nu(V,R) - \int_{\partial V[R]} d^c \log \|z\|^2 \wedge \omega^{k-1} \right\}.$$

For very small r, V is approximately its tangent cone V_0 whose equations are the homogeneous terms of lowest degree in the defining equations of V. The corresponding projective variety $P(V_0) \subset P^{n-1}$ satisfies

$$\text{degree}(P(V_0)) = \text{mult}_0(V) = \int_{P(V_0)} \omega^{k-1}$$

by the Wirtinger theorem again. In each line $\xi \in P^{n-1}$, $d^c \log \|z\|^2$ is the angular measure $d(\theta_\xi)$ where $r_\xi e^{i\theta_\xi}$ are polar coordinates in ξ. Thus

$$\lim_{r \to 0} \int_{\partial V[r]} d^c \log \|z\|^2 \wedge \omega^{k-1} = \text{degree}(V_0),$$

which proves our contention. Q.E.D.

The above proof gives the beautiful formula

$$(1.5) \quad \begin{cases} C_k \dfrac{\text{vol}\,(V[r])}{r^{2k}} = \nu(V,r) \\[2ex] \text{where} \\[2ex] \nu(V,r) = \text{mult}_0(V) + \{\text{volume in } P^{n-1} \text{ of the set of} \\ \qquad\qquad \text{lines } \overrightarrow{0z},\, 0 \neq z \in V[r]\}\,. \end{cases}$$

The function $\nu(V,r)$ is the basic growth indicator of an entire analytic set. For reasons arising from Jensen's theorem, we shall also have occasion to use the *counting function*

$$N(V,r) = \int_0^r \{\nu(V,\rho) - \nu(V,0)\} \frac{d\rho}{\rho} + \nu(V,0)\,\log r \,.$$

In case $V = \{a_1, a_2, \cdots\}$ is zero-dimensional, $\nu(V,r) = n(V,r)$ is the number of points in $B[r]$ and an integration by parts gives, assuming $\{0\} \notin V$,

$$N(V,r) = \int_0^r n(V,\rho) \frac{d\rho}{\rho} = \sum_{\|a_\mu\| \leq r} \log \frac{r}{\|a_\mu\|} \,.$$

In general, from (1.4) and (1.5) we observe the properties:

If $B[z,r]$ is the ball of radius r and center z, then the lower bound

$$(1.6) \qquad \text{vol}\,(V \cap B[z,r]) \geqq C_k r^{2k} \qquad (z \in V)$$

on the volume of V is valid;

If $V \subset C^n$ is a hypersurface ($k = n-1$), $\{0\} \notin V$, and if for each line $\xi \in P^{n-1}$ passing through the origin we let $n(V, \xi, r)$ be the number of points of $V[r] \cap \xi$, then

$$(1.7) \qquad\qquad \nu(V,r) = \int n(V, \xi, r)\, d\xi$$

is the average of $n(V, \xi, r)$ over all lines ξ.

(PROOF. $\omega^{n-1} = d\xi$ is the invariant density on P^{n-1}.)

If $V \subset C^n$ is an algebraic set of degree d, then

(1.8) $$\lim_{r \to \infty} \nu(V, r) = d$$

(PROOF: We may assume that $\{0\} \not\in V$. The projection of V on the hyperplane P^{n-1} at infinity is the set of lines $\overrightarrow{0z}$ $(z \in V)$, and is an algebraic variety of degree d in P^{n-1}.)

The converse of (1.8) is the important

THEOREM OF STOLL: V is algebraic if, and only if,

(1.9) $$\nu(V, r) = 0(1) .$$

We shall prove that

$$\text{``}\nu(V, r) \leq d \text{ implies V algebraic''}$$

in case V is a hypersurface. As with most questions involving higher codimension, the general situation is at present considerably more difficult.

For simplicity we assume that the origin does not lie on V. The residual mapping

$$\begin{cases} \pi : V \to P^{n-1} \\ \pi(z) = \overrightarrow{0z} \end{cases}$$

is equidimensional and $\nu(V, r)$ is the volume of $\pi(V[r])$. By (1.7), on the average V meets each line ξ through the origin in at most d points. What we shall prove is that this is true for every line ξ. If this has been done, then an easy argument (cf. the proof of (1.27) for divisors in C^n) shows that V is algebraic of degree $\leq d$.

For each point $z \neq 0$ in C^n we denote by L_z the line $\overrightarrow{0z}$, and by

$$n(V, z, r) = \#(V \cap L_z \cap B[\|z\|r])$$

the number of points of intersection of V with the line L_z in the ball of radius $\|z\|r$. We consider the logarithmic means

$$N(V,r) = \int_0^r \nu(V,\rho)\, \frac{d\rho}{\rho}$$

$$N(V,z,r) = \int_0^r n(V,z,\rho)\, \frac{d\rho}{\rho}\ .$$

It is standard (Cousin II problem in C^n) that there is an entire holomorphic function $f \epsilon \mathcal{O}(C^n)$ whose zero set is V and with $f(0) = 1$. By Jensen's formula (1.1),

$$N(V,z,r) = \frac{1}{2\pi} \int_0^{2\pi} \log|f(re^{i\theta}z)|\, d\theta\ .$$

It follows that $N(V,z,r)$ is a *plurisubharmonic function* of $z \epsilon C^n$. Such functions $\psi(z)$ satisfy the sub-mean-value principle

$$\psi(z) \leq \frac{C_n}{\sigma^{2n}} \int_{B[z,\sigma]} \psi(w)\Phi(w)$$

where $\Phi(w)$ is Euclidean measure. If furthermore $\psi \geq 0$, then

$$\psi(z) \leq \frac{C}{\|z\|^{2n}} \int_{B[2\|z\|]} \psi(w)\Phi(w)\ .$$

Applying this to $N(V,z,r)$ and taking $\|z\| = 1$ and using (1.7) gives the estimate

(1.10) $N(V,z,r) \leq C(N(V,2r))\ ,$

valid for general analytic hypersurfaces V in C^n.

Now it is clear that $n(V,z,r)$ is $0(1)$ if, and only if, $N(V,z,r)$ is $0(\log r)$ and similarly for $\nu(V,r)$ and $N(V,r)$. When combined with (1.10), we are done.

(c) *Order functions for entire holomorphic mappings*

By definition, an *entire holomorphic mapping* is a holomorphic mapping

$$f : C^n \to M$$

into a compact, complex manifold M. If $\dim_C M = n$, we shall say that f
is *equi-dimensional*, and in this case f is *non-degenerate*, when the
Jacobian determinant is not identically zero. The special case

$$f : C \to P^1$$

is just an entire meromorphic function, frequently written as $w = f(z)$.

We shall measure growth of an entire holomorphic mapping relative to
a positive holomorphic line bundle $L \to M$. Let η be the *Chern form* of
this line bundle relative to some fiber metric. If s is a non-vanishing
holomorphic section of L over an open set on M, then by definition

(1.11) $$\eta = dd^C \log \frac{1}{|s|^2} \; .$$

A different choice of metric $|s|'^2$ gives a new Chern form η' related to
η by

(1.12) $$\eta' = \eta + dd^C \sigma$$

where $\sigma = \log(|s|'^2/|s|^2)$ is a C^∞ function on M. In particular, the
deRham cohomology class

$$[\eta] \in H^2_{DR}(M, R)$$

is well-defined. The potential theory on M which we shall require is
contained in the converse to (1.12):

(1.13) LEMMA. *Let ψ be a real, closed (1,1) form such that $[\psi] = [\eta]$
in $H^2_{DR}(M,R)$. Then ψ is the Chern form of some metric in $L \to M$.*

PROOF. Since M has a positive line bundle, it carries a Kähler metric
and we let Δ_d and Δ_{d^C} be the Laplace-Beltrami operators of d and

d^C relative to this metric. By a standard Kähler identity

$$\Delta_d = \Delta_{d^c} \, .$$

Thus the Green's operators also satisfy

$$G_d = G_{d^c} \, .$$

Since $\eta - \psi$ is zero in cohomology,

$$\eta - \psi = \Delta_d \, G_d(\eta, \psi)$$

as follows from the Hodge decomposition. Similarly

$$\eta - \psi = \Delta_{d^c} G_{d^c} (\eta - \psi) \, .$$

Combining these with $d(\eta - \psi) = 0 = d^c(\eta - \psi)$ gives the relation

$$\eta - \psi = dd^c(d^* d^{c^*} G_d G_{d^c}(\eta - \psi)) = dd^c \sigma \, .$$

If we multiply the given metric in $L \to M$ by $e^{-\sigma}$, then the new Chern form is ψ. \hfill Q.E.D.

For an entire holomorphic mapping $f : C^n \to M$, we define

$$t_f(L, r) = \int_{B[r]} f^* \eta \wedge \omega^{n-1}$$

$$T_f(L, r) = \int_0^r t_f(L, \rho) \frac{d\rho}{\rho} \, .$$

$T_f(L, r)$ is called the *order function* of f relative to the line bundle $L \to M$. Interpreting ω^{n-1} as the standard measure $d\xi$ on the projective space P^{n-1} of lines ξ through the origin in C^n,

$$t_f(L, r) = \int_{P^{n-1}} \left(\int_{\xi[r]} (f^* \eta) \right) d\xi, \quad (\xi[r] = \xi \cap B[r]) \, .$$

It follows that

$$(1.14) \qquad T_f(L, r) = \int_{\mathbf{P}^{n-1}} T_f(L, \xi, r) \, d\xi$$

is the average of the order functions of f restricted to lines ξ.

Using (1.12) and Stokes' theorem together with

$$d^c = \frac{1}{4\pi} r \frac{\partial}{\partial r} \otimes d\theta - \frac{1}{4\pi} \frac{1}{r} \frac{\partial}{\partial \theta} \otimes dr \, ,$$

a different choice of metrics gives a new order function

$$T_f'(L, \xi, r) = T_f(L, \xi, r) + \int_0^r \left(\int_{\xi[\rho]} dd^c \sigma_f \right) \frac{d\rho}{\rho} \qquad (\sigma_f = f^* \sigma)$$

$$= T_f(L, \xi, r) + \int_0^r \left(\int_{\partial \xi[\rho]} d^c \sigma_f \right) \frac{d\rho}{\rho}$$

$$= T_f(L, \xi, r) + \int_0^r \frac{1}{4\pi} \rho \frac{\partial}{\partial \rho} \left(\int_{\partial \xi[\rho]} \sigma_f \, d\theta \right) \frac{d\rho}{\rho}$$

$$= T_f(L, \xi, r) + \left[\frac{1}{4\pi} \int_{\partial \xi[r]} \sigma_f \, d\theta - \frac{1}{2} \sigma_f(0) \right]$$

$$= T_f(L, \xi, r) + 0(1) \, ,$$

because σ is bounded on M. Since $L \to M$ has been assumed to be positive, there is some metric for which η is positive, and consequently

$$T_f(L, r) \geqq C \log r$$

unless f is constant, a case which we exclude. It follows from these two observations that the growth of the order function is intrinsically defined up to a relatively insignificant $0(1)$ term.

Suppose now that $D \in |L|$ is the divisor of a holomorphic section $s \in \mathcal{O}(M, L)$, and $f : \mathbf{C}^n \to M$ is an entire holomorphic mapping whose image does not lie in D. Then $D_f = f^{-1}(D)$ is an effective divisor in C^n, which may be described as the zero locus of the section f^*s of $f^*L \to C^n$. The *First Main Theorem* (F.M.T.) gives the basic relation between the growth functions

$$\begin{cases} N_f(D, r) = N(D_f, r) \\ \text{and} \\ T_f(L, r) \ . \end{cases}$$

To state it, we may multiply s by a non-zero constant to assume that the length $|s|^2 \leq 1$, and then define the *proximity form*

$$m_f(D, r) = \frac{1}{2} \int_{\partial B[r]} \left(\log \frac{1}{|f^*s|^2} \right) \Sigma \geq 0$$

where

$$\Sigma = d^c \log \|z\|^2 \wedge (dd^c \log \|z\|^2)^{n-1}$$

is the unique, closed $2n-1$ form on $C^n - \{0\}$ which is invariant under unitary transformations and has integral one over a sphere of any radius. When $n = 1$ and $z = re^{i\theta}$,

$$\Sigma = \frac{1}{2\pi} d\theta \ ,$$

and in general

$$\Sigma = \frac{1}{2\pi} d\theta_\xi \wedge d\xi$$

where $d\theta_\xi$ is angular measure in the line $\xi \in P^{n-1}$. With the obvious notation,

$$m_f(D, r) = \int_\xi m_f(D, \xi, r) d\xi \ .$$

The F.M.T. is the formula

(1.15) $$N_f(D, r) + m_f(D, r) = T_f(L, r) + 0(1)$$

where the $0(1)$ is independent of $D \in |L|$.

PROOF. Since all terms appearing in (1.15) are averages over the lines $\xi \, \epsilon \, P^{n-1}$ of the corresponding 1-variable quantities, it will suffice to prove the result in the special case $n = 1$.

If f^*s has no zeroes, then by the usual computation

$$T_f(L, r) = \int_0^r \left(\int_{\Delta(\rho)} dd^c \log \frac{1}{|f^*s|^2} \right) \frac{d\rho}{\rho}$$

$$= \int_0^r \left(\int_{\partial\Delta(\rho)} d^c \log \frac{1}{|f^*s|^2} \right) \frac{d\rho}{\rho} \qquad \text{(Stokes')}$$

$$= \int_0^r \frac{1}{2} \, \rho \, \frac{\rho}{d\rho} \left(\frac{\rho}{2\pi} \int \log \frac{1}{|f^*s|^2} \, d\theta \right) \frac{d\rho}{\rho} \qquad \text{(formula for } d^c)$$

$$= m_f(D, r) + C \, .$$

In the general case, we may write

$$f^*s = h \cdot s'$$

on $\Delta(r+\epsilon)$ where s' is a non-zero holomorphic section of f^*L and h is a holomorphic function having the same zeroes as f^*s there. The F.M.T. (1.15) is then the sum of the formula just proved, applied this time to s' instead of f^*s, and the Jensen theorem (1.1) for h.   Q.E.D.

COROLLARY (Nevalinna inequality):

(1.16) $N_f(D, r) \leqq T_f(L, r) + 0(1)$

In general, the proximity form $m_f(D, r)$ is large when $f(\partial B[r])$ is on the mean close to D, so that in some sense the left hand side of (1.15) measures the *total attraction* of $f(z)$ to the divisor D in the ball of radius r. The F.M.T. says that this total attraction is essentially independent of the particular divisor.

An important special case of an entire holomorphic mapping is when $M = P^m$ is complex projective space and L is the hyperplane line bundle H. Denoting by $Z = [z_0, \cdots, z_m]$ the homogeneous coordinates of a point in $\mathbf{P}^m$ and by $A = [a_0, \cdots, a_m]$ points in the dual projective space $\mathbf{P}^{m*}$ of hyperplanes in $\mathbf{P}^m$, the divisors $A \in |H| = \mathbf{P}^{m*}$ are linear spaces given by an equation

$$<A, Z> = a_0 z_0 + \cdots + a_m z_m = 0 .$$

The length function of a section defining A may be taken to be

$$u(A, Z) = \frac{|<A, Z>|}{\|A\| \ \|Z\|} .$$

The subsequent Chern form

$$\eta = dd^c \log \frac{1}{u(A, Z)^2} = dd^c \log \|Z\|^2$$

is the standard Kähler form (*Fubini-Study metric*) on projective space.

The unitary group $\mathbf{U}_{m+1}$ acts on $\mathbf{P}^m$ and $\mathbf{P}^{m*}$, and we denote by dA the unique invariant measure on $\mathbf{P}^{m*}$ with total volume one. Equivalently, if $v(A)$ is an integrable function on $\mathbf{P}^{m*}$, then

$$\int_{\mathbf{P}^{m*}} v(A) \, dA = \int_{T \in \mathbf{U}_{m+1}} v(T \cdot A_0) \, dT$$

where dT is the invariant measure on the unitary group and A_0 is a fixed point in $\mathbf{P}^{m*}$ (by invariance, it doesn't matter which A_0 we select). Perhaps the best interpretation of the order function of $f : \mathbf{C}^n \to \mathbf{P}^m$ is that furnished by *Crofton's formula*:

$$(1.17) \qquad T_f(H, r) = \int_{A \in \mathbf{P}^{m*}} N_f(A, r) \, dA$$

expressing the order function as the average of the counting functions $N_f(A, r)$.

PROOF. Using unitary invariance of the inner product and of the measure dT,

$$\int_{A \in P^{m*}} \log u(A, Z) \, dA = \int_{T \in U_{m+1}} \log u(TA_0, Z) \, dT$$

$$= \int_{T \in U_{m+1}} \log u(STA_0, SZ) \, dT$$

$$= \int_{T \in U_{m+1}} \log u(STA_0, SZ) \, d(ST)$$

$$= \int_{A \in P^{m*}} \log u(A, SZ) \, dA$$

for any fixed $S \in U_{m+1}$. It follows that

$$\int_{A \in P^{m*}} m_f(A, r) \, dA$$

is a constant C_0. Averaging (1.15) gives

$$\int_{A \in P^{m*}} N_f(A, r) \, dA = T_f(H, r) + C \ ,$$

and $C = 0$ by letting $r \to 0$. Q.E.D.

The tension between the simultaneous relations

$$\begin{cases} N_f(A, r) \leqq T_f(H, r) + 0(1) \\ \\ \int N_f(A, r) \, dA = T_f(H, r) \end{cases}$$

suggests the deeper aspects of the theory. For example, it follows immediately that the image $f(C^n)$ cannot omit an open set U of hyperplanes, since such a U would have positive measure. This is the *Liouville theorem*, derived here by purely integral methods.

As an application of (1.16) and (1.17), we shall prove:

The mapping f *is rational if, and only if,*

$$T_f(L, r) = O(\log r) \ .$$

Before giving the argument, some explanation may be in order. Since the line bundle $L \to M$ is assumed to be positive, M has uniquely the structure of a projective algebraic variety. More precisely, by a famous theorem of Kodaira, some high power $L^k = \underbrace{L \otimes \cdots \otimes L}_{k\text{-times}}$ of the bundle L will have enough holomorphic sections to induce a projective embedding

$$\left\{ \begin{array}{l} j : M \to \mathbf{P}^N \\[2mm] \text{with} \\[2mm] j^*(H) = L^k \ , \end{array} \right.$$

and where $|H|$ is the linear system of hyperplanes in $\mathbf{P}^N$. Since

$$T_f(L^k, r) = kT_f(L, r) \ ,$$

we may as well assume that $M = \mathbf{P}^m$ and $L = H$ is the standard line bundle. Then f is rational if, and only if, the functions $f^*(\mathfrak{z}_i/\mathfrak{z}_0)$ $(i = 1, \cdots, m)$ are rational functions on $\mathbf{C}^n$. Now an entire meromorphic function $g(z) \, (z \in \mathbf{C})$ is rational of degree d if, and only if, the equation

$$g(z) = a$$

has at most d solutions for all points $a \in \mathbf{P}^1$. Similarly, an entire meromorphic function $g(z) \, (z \in \mathbf{C}^n)$ is rational of degree d if, and only if, all level sets $g(z) = a$ are algebraic hypersurfaces of degree $\leq d$. In the question at hand, these level sets are inverse images $f^{-1}(A)$ of linear hyperplanes in $\mathbf{P}^m$.

Thus, if f is rational of degree d, then $N_f(A, r) \leq d \log r + C$ for all $A \in \mathbf{P}^{m*}$, and by Crofton's formula (1.17)

$$T_f(H, r) \leq d \log r + C.$$

Conversely, if this estimate on the order function is valid, then by
(1.16)
$$N_f(A, r) \leq d \log r + C$$

for all A and by Stoll's theorem (1.9) the mapping f is rational of
degree $\leq d$.   Q.E.D.

(d) *Classical indicators of orders of growth*

Working backwards from a historical viewpoint, we shall discuss some
classical measurements of growth and give a few applications. For an
entire meromorphic function
$$f : \mathbf{C} \to \mathbf{P}^1$$

and hyperplane (= point) line bundle, our order function reduces to the
Ahlfors-Shimizu characteristic function

$$T_f(H, r) = \int_0^r \left(\int_{\Delta(\rho)} f^*\eta \right) \frac{d\rho}{\rho} = \int_0^r A_f(\rho) \frac{d\rho}{\rho}$$

where

$$A(\rho) = \frac{\sqrt{-1}}{2\pi} \int_{\Delta(\rho)} \frac{|f'(z)|^2 \, dz \wedge d\bar{z}}{(1 + |f(z)|^2)^2}$$

is the *spherical area* of the image $f(\Delta(\rho))$. This order function was used
as an alternative to the *Nevanlinna characteristic function*

$$T_f(r) = N_f(\infty, r) + \frac{1}{2\pi} \int_0^{2\pi} \log^+ |f(re^{i\theta})| \, d\theta$$

where

$$\log^+ a = \max (0, \log a) .$$

Thus, $T_f(r)$ measures the total attraction of f on $\Delta(r)$ to the point at
infinity on $\mathbf{P}^1 = \mathbf{C} \cup \{\infty\}$. Using

$$\log a = \log^+ a - \log^+ \frac{1}{a} \, ,$$

Jensen's formula (1.2) may be written as

$$(1.18) \qquad N_f(a, r) + \frac{1}{2\pi} \int_0^{2\pi} \log^+ \frac{1}{|f(re^{i\theta}) - a|} \, d\theta = T_f(r) + O(1) \, ,$$

expressing the symmetry of total attraction of f to all points $a \, \epsilon \, P^1$. The formula (1.18) was the F.M.T. in Nevanlinna's original work. This seemingly innocuous splitting of Jensen's formula to give (1.18) and subsequent interpretation provided the key to the theory.

If we use the inequalities

$$\log^+ |f(re^{i\theta})| \leq \log (1 + (f(re^{i\theta})|^2)^{\frac{1}{2}} \leq \log^+ |f(re^{i\theta})| + \log 2$$

in the F.M.T. (1.15)

$$T_f(H, r) = N_f(\infty, r) + \frac{1}{4\pi} \int_0^{2\pi} \log (1 + |f(re^{i\theta})|^2) \, d\theta$$

for the Ahlfors-Shimizu characteristic function and in the definition of $T_f(r)$, we find that

$$T_f(r) \leq T_f(H, r) \leq T_f(r) + \log 2 \, ,$$

so that the two order functions are essentially equivalent. The Nevanlinna characteristic function has the substantial advantage that algebraic relations such as

$$(1.19) \qquad \begin{cases} T_{f+g}(r) \leq T_f(r) + T_g(r) + O(1) \\ T_{fg}(r) \leq T_f(r) + T_g(r) \\ T_{1/f}(r) = T_f(r) + O(1) \end{cases}$$

are apparent.

The relation between $T_f(r)$ and $T_{f'}(r)$ is considerably more subtle and in some sense may be said to hold the key to the defect relations. Indeed, R. Nevanlinna's main technical tool was an estimate on

$$\frac{1}{2\pi} \int_0^{2\pi} \log^+ \left| \frac{f'(re^{i\theta})}{f(re^{i\theta})} \right| d\theta$$

in terms of $\underline{\log} \ T_f(r)$. We shall return to these questions in Chapters 2 and 3.

For the moment we wish to discuss the relation between $T_f(r)$ and the *maximum modulus indicator*

$$M_f(r) = \max_{|z| \leqq r} \log |f(z)|$$

of an entire *holomorphic* function $f(z)$. Certainly $M_f(r)$ is the most obvious, and also most classical, growth indicator. We shall show that $M_f(r)$ and $T_f(r)$ grow at essentially the same rate by using the *Poisson-Jensen-Nevanlinna formula* for meromorphic functions

(1.20)
$$\log |f(z)| = \frac{1}{2\pi} \int_0^{2\pi} \log |f(re^{i\theta})| \, P(re^{i\theta}, z) \, d\theta$$
$$+ \sum_{|b_\nu| < r} g(z, b_\nu) - \sum_{|a_\mu| < r} g(z, a_\mu)$$

where

$$P(re^{i\theta}, \rho e^{i\phi}) = \frac{r^2 - \rho^2}{r^2 + \rho^2 - 2r\rho \cos(\phi - \theta)}$$

is the *Poisson kernel*,

$$g(z, w) = \log \left| \frac{r^2 - w\bar{z}}{r(z-w)} \right| \geqq 0$$

is a *Green's function*, and $\{a_\mu\}$, $\{b_\nu\}$ are the respective zeroes and poles of $f(z)$ in $\Delta(r)$. To prove (1.19) we write

$$N_f(0, r) = \int_0^r n_f(0, \rho) \frac{d\rho}{\rho} = \sum_\mu \log \frac{r}{|a_\mu|}$$

by an integration by parts (assuming $f(0) \neq 0, \infty$), and similarly for $N_f(\infty, r)$. Then (1.20) results from the usual Jensen's theorem (1.2) by making a standard linear fractional transformation in $\Delta(r)$ sending z to 0.

If case $f(z)$ is entire holomorphic, we have

(1.21) $T_f(r) \leq M_f(r) \leq \left(\dfrac{a+1}{a-1}\right) T_f(ar)$ $(a > 1)$.

PROOF. The first inequality is obvious, and the second follows from (1.20) and

$$|P(a\,re^{i\theta}, z)| \leq \frac{a+1}{a-1}$$

valid for $|z| \leq r$ and $a > 1$.

Thus, $T_f(r)$ gives a measure of growth for meromorphic functions generalizing the maximum modulus of a holomorphic function.

As a non-obvious (at least to me) application of (1.21), we have the result: *If* $f(z), g(z)$ *are entire holomorphic functions such that* f/g *is also holomorphic, then*

(1.22) $M_{f/g}(r) \leq \left(\dfrac{a+1}{a-1}\right)\{M_f(ar) + M_g(ar)\} + C$ $(a > 1)$

We shall use (1.22) to complete the proof of Picard's theorem begun in the introduction to this chapter.

First we observe that if $f(z)$ is a holomorphic function, then Cauchy's formula

$$f'(z) = \frac{1}{2\pi\sqrt{-1}} \int_{|w|=R} \frac{f(w)\,dw}{(w-z)^2} \qquad (|z| < R)$$

gives an estimate

(1.23) $M_{f'}(r) \leq C\,M_f(ar)$ $(a > 1)$.

Next, we recall the *Borel-Caratheodory lemma*

(1.24) $|f(z)| \leq C_a \max_{|w| \leq ar} |Re\,f(w)|$ $(|z| < r)$

which is proved as follows: Applying (1.20) to $F = e^f$ gives for $|z| < r$

$$\operatorname{Re} f(z) = \frac{1}{2\pi} \int_{|w|=r} \operatorname{Re} f(w) \operatorname{Re}\left(\frac{w+z}{w-z}\right) d(\arg w) .$$

Taking harmonic conjugates we find

$$\operatorname{Im} f(z) = \frac{1}{2\pi} \int_{|w|=r} \operatorname{Re} f(w) \operatorname{Im}\left(\frac{w+z}{w-z}\right) d(\arg w) .$$

The Borel-Caratheodory lemma follows by adding these and making the obvious estimate.

Now to the Picard theorem. As discussed in (a) of this chapter, an entire meromorphic function omitting three values on P^1 leads to a linear relation

$$e^g + e^h = 1$$

between exponentials of entire holomorphic functions g and h. Differentiating and solving gives

$$\begin{cases} e^g = \dfrac{h'}{h'-g'} \\[2ex] e^h = \dfrac{g'}{g'-h'} \end{cases} .$$

This holds unless $h' \equiv g'$, in which case the original map is constant. Setting

$$S(r) = M_g(r) + M_h(r) ,$$

if we combine (1.22)-(1.24) we obtain

$$S(r) \leqq \log S(ar) + O\left(\log \frac{1}{a-1}\right) + C .$$

To show that this leads to a contradiction, we use the following lemma of Borel which was devised for just this purpose:

LEMMA. Let $S(r) > 0$ be continuous and monotonically increasing for $r_0 \leqq r < \infty$. Then

$$S\left(r + \frac{1}{S(r)}\right) < 2S(r)$$

outside an exceptional set E *of finite total measure.*

PROOF. Assuming E is non-empty, set $r_1 \epsilon E$, $S_1 = S(r_1) > 0$ and $\Delta_1 = \frac{1}{S_1}$. Inductively define r_k, $S_k = S(r_k)$, $\Delta_k = \frac{1}{S_k}$ by

$$r_{k+1} = \min\{r \epsilon E : r_k + \Delta_k \leqq r\} .$$

Then

$$\sum_k \Delta_k = \sum_k \frac{1}{S_k} \leq \frac{1}{S_1}\left(\sum_k \frac{1}{2^k}\right) = \frac{1}{S_1} ,$$

and E is covered by the Δ_k.   Q.E.D.

To complete the proof of Picard's theorem, we determine a by

$$ar = r + \frac{1}{S(r)} ,$$

or

$$\frac{1}{a-1} = S(r) .$$

Then the inequality on $S(r)$ becomes

$$S(r) \leqq \log S\left(r + \frac{1}{S(r)}\right) + C' \log S(r) + C .$$

Comparing this with Borel's lemma yields the desired contradiction.

(e) *Entire functions and varieties of finite order*

We shall conclude this first chapter by discussing analytic functions and varieties of finite order. Thus far, these are the most important entire transcendental quantities.

Given a positive increasing function $\phi(r)$, if the upper limit

$$\varlimsup_{r \to \infty} \frac{\log \phi(r)}{\log r} = \lambda$$

is finite, then $\phi(r)$ is said to have finite order λ. This is equivalent to either of the following assertions:

Given $\varepsilon > 0$, $\phi(r) < r^{\lambda + \varepsilon}$ for all
large r, while $\phi(r_n) > r_n^{\lambda - \varepsilon}$ for some sequence $r_n \to \infty$;

The integral $\displaystyle\int_1^\infty \frac{\phi(t) dt}{t^{\mu + 1}}$ is convergent for $\mu > \lambda$ and

divergent for $\mu < \lambda$.

PROOF. If $\phi(r) < r^{\lambda + \varepsilon}$ for large r, then $\int_1^\infty \frac{\phi(t) dt}{t^{\mu + 1}}$ evidently converges

for $\mu > \lambda$. If conversely this integral converges, then for r sufficiently
large

$$\frac{1}{\mu} > \int_r^\infty \frac{\phi(t) dt}{t^{\mu + 1}} \geqq \frac{\phi(r)}{\mu r^\mu} .$$

If we set

$$\Phi(r) = \int_1^r \frac{\phi(t) dt}{t}$$

and use integration by parts

$$\mu \int_1^r \frac{\Phi(t) dt}{t^{\mu + 1}} + \frac{\Phi(r)}{r^\mu} = \int_1^r \frac{\phi(t) dt}{t^{\mu + 1}} + C ,$$

we find that:

$\Phi(r)$ has finite order r if, and only if, this is true of $\phi(r)$.

The following definitions are the obvious ones: An entire analytic
set $V \subset C^n$ has finite order λ if this is true for $\nu(V, r)$, or equivalently
for the counting function $N(V, r)$. An entire holomorphic mapping $f : C^n \to M$
has finite order λ if this is true for the order function $T_f(L, r)$. (Remark:
Since for any two positive line bundles L and L',

$$\frac{1}{C} T_f(L, r) \leqq T_f(L', r) \leqq C T_f(L, r)$$

for a suitable $C > 0$, the finite order condition is independent of the positive line bundle.)

For entire meromorphic functions we may use either the Ahlfors-Shimizu or Nevanlinna characteristic functions. For entire holomorphic functions we may additionally use the maximum modulus indicator. The order is the same in all cases.

For an entire holomorphic function $f(z)$ of finite order λ, the zero locus $D = \{a_1, a_2, \cdots\}$ has finite order $\leq \lambda$. This follows from Jensen's theorem (1.1), and here is a proof based on the maximum principle: To begin with, assuming $\{0\} \notin D$

$$N(D, r) = \int_0^r n(D, \rho) \frac{d\rho}{\rho} = \sum_{\mu=1}^{n(D,r)} \log \frac{r}{|a_\mu|}$$

using an integration by parts. Consider the analytic function

$$g(z) = \frac{f(z)}{\displaystyle\prod_{\mu=1}^{n(D,r)} (z - a_\mu)}$$

in the larger disc $|z| \leq ar$. By the maximum principle,

$$\frac{|f(0)|}{\displaystyle\prod_\mu |a_\mu|} = |g(0)| \leq \frac{\displaystyle\max_{|z|=ar} |f(z)|}{\displaystyle\min_{|z|=ar} \prod |z - a_\mu|} \leq \frac{e^{M_f(ar)}}{(a-1) r^{n(D,r)}} \ .$$

Taking logarithms gives

$$\log |f(0)| + \sum_{\mu=1}^{n(D,r)} \log \frac{r}{|a_\mu|} \leq M_f(ar) + \log \left(\frac{1}{a-1}\right) .$$

Conversely, given $D = \{a_\mu\}$ of finite order λ in C, we may find an entire function f of order λ whose zero set is D. For this one considers the *Weierstrass primary factors*

$$E(u, q) = (1-u)e^{u+u^2/2 + \cdots + u^q/q} .$$

Elementary considerations give the inequalities

(1.25)
$$\begin{cases} \log |E(u,q)| \leq \log(1+(u)) < |u| \\[2ex] \log |E(u,q)| \leq C\dfrac{|u|^{q+1}}{1+|u|} \leq C|u|^{q+1} \qquad (q \geq 1) . \end{cases}$$

Choosing a smallest q such that the series $\sum\limits_{\mu} \dfrac{1}{|a_\mu|^{q+1}}$ converges, the infinite product

$$\prod_{\mu} E\left(\frac{z}{a_\mu}, q\right)$$

converges normally to an entire function $f(z)$ having D as its zero locus

By what was proved earlier, $f(z)$ has order at least λ. On the other hand, using integration by parts

$$\sum_{|a_\mu|<r} \frac{1}{|a_\mu|^\rho} = \int_0^r \frac{dn(D,t)}{t^\rho} = \frac{n(D,r)}{r^\rho} + \frac{1}{\rho}\int_0^r \frac{n(D,t)dt}{t^{\rho+1}} .$$

Taking $\rho = q+1$ it follows that $f(z)$ has finite order $\lambda \leq q+1$. To obtain a more precise estimate, in case $q \geq 1$ using (1.25), we have

(1.26)
$$M_f(r) \leq C \sum \frac{r^{q+1}}{|a_\mu|^q(|a_\mu|+r)} = Cr^{q+1}\int_0^\infty \frac{dn(D,t)}{t^q(t+r)}$$

$$= Cr^{q+1}\int_0^\infty \frac{(qr+(q+1)t)n(D,t)dt}{t^{q+1}(t+r)^2}$$

$$\leq C(q+1)r^{q+1}\int_0^\infty \frac{n(D,t)dt}{t^{q+1}(t+r)} ,$$

or

$$M_f(r) \leq C'r^q\left[\int_0^r \frac{n(D,t)dt}{t^{q+1}} + r\int_0^\infty \frac{n(D,t)dt}{t^{q+2}}\right] .$$

A similar argument gives the same estimate when $q = 0$. Taking any $\mu > \lambda$ so that $n(D, t) < C'' t^\mu$, we find from (1.26) that

$$M_f(r) < C''' r^\mu .$$

Thus $f(z)$ has finite order exactly equal to λ, and we have proved one-half of the Hadamard factorization theorem:

(1.27) *Given* $D = \{a_\mu\}$ *a divisor of finite order* λ *in* C, *there exists an entire holomorphic function* $f(z)$ *of order* λ *having* D *as its zero locus. If* $g(z)$ *is another such function, then*

$$g(z) = e^{p(z)} f(z)$$

where $p(z)$ *is a polynomial of degree* $\leq [\lambda]$.

The uniqueness follows by writing $g/f = e^p$ for some entire function h, using (1.22) to deduce that $\max_{|z| \leq r} \operatorname{Re} p(z) = O(r^{\lambda + \varepsilon})$ for every $\varepsilon > 0$, and applying the Borel-Caratheodory lemma (1.24).

Using the one variable result and semi-continuity principle (1.10), we shall prove the same theorem (1.27) for divisors of finite order in C^n, this result being due to Lelong and Stoll. To begin with, if $f(z)$ is an entire holomorphic function of $z = (z_1, \cdots, z_n) \in C^n$ such that

$$M_f(r) = \max_{\|z\| \leq r} \log |f(z)|$$

has finite order λ, then the zero locus D of $f(z)$ has finite order $\leq \lambda$ by the one-variable result and averaging formula (1.7). If D has order exactly equal to λ and $g(z)$ also defines D, then $g = e^p f$ where $p(z)$ is a polynomial of degree $\leq [\lambda]$ by the one variable result applied to the lines through the origin. It remains to prove existence.

For convenience we assume that $\{0\} \notin D$. By (1.10) the intersections $D_\xi = D \cap \xi$ all have finite order $\leq \lambda$. We shall use the Weierstrass products on each line ξ and patch together by uniqueness. Here is the proof in detail for $n = 2$, the general case being similar.

We cover $C^2 - \{0, 0\}$ by the two open sets $U_i = \{(z_1, z_2) : z_i \neq 0\}$ ($i = 1, 2$). The picture is something like

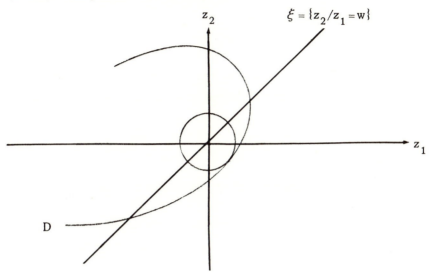

The holomorphic mapping

(1.28) $(v, w) \to (v, vw)$

maps C^2 onto $(C^2 - \{z_1 = 0\}) \cup \{(0, 0)\}$, and has the inverse

$$(z_1, z_2) \to (z_1, z_2/z_1)$$

on U_1. For fixed w, the mapping (1.28) gives a parametric representation of a line L_w. The boundary of the ball of radius r meets L_w in the circle

$$|v| = \frac{r}{(1 + |w|^2)^{\frac{1}{2}}}.$$

The intersection

$$D \cdot L_w = v_1(w) + v_2(w) + \cdots$$

is a divisor in the v-plane of finite order $\leq \lambda$. The points $v_i(w)$ are algebroid functions of w; i.e. they locally vary in a holomorphic manner with w, except for branching.

Setting $q = [\lambda] + 1$, the product

$$\prod_{\mu} E\left(\frac{v}{v_{\mu}(w)}, q\right)$$

converges normally in (v, w) to a holomorphic function having finite order $\leq \lambda$ on each line $w = $ constant. In fact, if we assume for simplicity that

$$N(D, r) \leq Cr^{\lambda},$$

then by (1.10)

$$N(D \cdot L_w, R) \leq C'(R^{\lambda}|w|^{\lambda})$$

where C' is independent of w and $N(D \cdot L_w, R)$ is the logarithmically averaged counting function for the number of $v_i(w)$ satisfying $|v_i(w)| < R$. It follows from (1.26) that

$$\max_{|v| \leq R} \log |f(v, w)| \leq C''(R^{\lambda}|w|^{\lambda}).$$

The function

$$f_1(z_1, z_2) = f(z_1, z_2/z_1)$$

is defined in U_1 and satisfies

$$\max_{|z_1| \leq R} \log |f_1(z_1, z_2)| \leq C''\left(R^{\lambda}\left|\frac{z_2}{z_1}\right|^{\lambda}\right).$$

On each fixed line $\{z_2/z_1 = $ constant$\}$, f_1 has finite order $\leq \lambda$ and extends holomorphically across the origin. Moreover, for fixed z_2

$$\log |f_1(z_1, z_2)| = 0(|z_1|^{-\lambda})$$

as $|z_1| \to 0$. The picture is something like

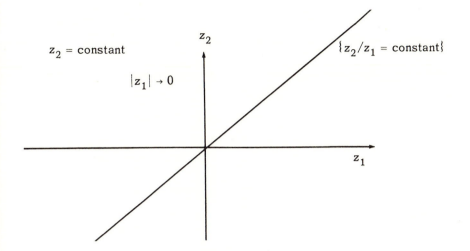

z_2 = constant

$|z_1| \to 0$

$\{z_2/z_1 = \text{constant}\}$

Similarly, we may construct f_2 on U_2. On the intersection $U_1 \cap U_2$,

$$f_1 = h_{12} f_2$$

where h_{12} is non-vanishing. Now if $h(z)$ is non-vanishing on the punctured plane $0 < |z| < \infty$, then $z^{-k} h(z)$ has a global logarithm where

$$k = \frac{1}{2\pi \sqrt{-1}} \int_{|z| = r} d \log h \,.$$

Similarly, $h_{12} z_1^{k_1} z_2^{k_2}$ has a global logarithm on $U_1 \cap U_2$. Multiplying f_1 by $z_1^{-k_1}$ and f_2 by $z_2^{k_2}$, we may assume that

$$f_1 = e^{p_{12}} f_2$$

on $U_1 \cup U_2$.

We now expand p_{12} in a Laurent series

$$p_{12} = \sum_{m,n = -\infty}^{+\infty} a_{mn} z_1^m z_2^n \,.$$

For fixed z_2,

$$\text{Re } p_{12} = 0(|z_1|^{-\lambda})$$

as $|z_1| \to 0$, and thus $a_{mn} = 0$ for $m < - [\lambda]$. Similarly $a_{mn} = 0$ for $m < - [\lambda]$. On any line $z_1 = \xi \cdot z_2 (\xi \neq 0)$, f_1 and f_2 have at most a pole at the origin. Thus $a_{mn} = 0$ for $m+n < 0$. Similarly, $a_{mn} = 0$ for $m > [\lambda]$, $n > [\lambda]$, or $m+n > [\lambda]$. Thus p_{12} is a finite Laurent series of total degree at most $[\lambda]$ in both positive and negative powers of z_1 and z_2. Setting

$$p_1 = - \sum_{n < 0} a_{mn} z_1^m z_2^n ,$$

$$p_{12} = - p_1 + p_2$$

where p_1 is holomorphic in U_1 and p_2 is holomorphic in U_2. Thus

$$e^{p_1} f_1 = e^{p_2} f_2 = f$$

gives a holomorphic function in $U_1 \cap U_2 = C^2 - \{0, 0\}$, and hence by Hartogs' theorem on all of C^2, a function which has zero divisor D. Moreover, f has finite order $\leqq \lambda$ by the growth properties of f_1, f_2, p_1, and p_2.   Q.E.D.

Bibliographical remarks for Chapter 1

The original proof of Stoll's theorem (1.9) appears in [38]. Interestingly enough, his argument uses Nevanlinna theory, but applied to the analytic variety V. A proof of (1.9) in local form was given by Bishop [5] with refinements by Shiffman [34]. The proof presented here for the case of hypersurfaces is one of the two found in [22]; the other one given there also uses value distribution theory on V and is similar in spirit to Stoll's original argument. Proposition (1.3) may be found in Lelong [30] and also in the paper by Stoll [38]. The multiplicity statement is due to Thie and Draper. A detailed discussion using blowing up, together with the original references, appears in Griffiths-King [22]. The general technique of reducing questions about an analytic function or analytic hypersurface in C^n to the one variable case seems to date back to Kneser [27], and has been used by Stoll [39] and Kujala [26] among others.

The potential-theoretic lemma 1.13 is due to Kodaira. The argument we have given assumes Hodge theory for compact manifolds and the standard Kähler identities, both of which are proved in Wells [43] or in the set of notes [14]. The order function $T_f(L, r)$ is a special case of that used by Bott and Chern [6]. Our First Main Theorem is a special case of the formalism in Stoll [37]. A general discussion of order functions appears in [10].

Crofton-type formulae expressing growth indicators as averages of counting functions are ubiquitous in the theory. For example, (1.7), (1.14), and (1.17) are all of this kind. The general philosophy comes from integral geometry, for which we refer to the book by Santalo [33]. A proof of Crofton's formula for holomorphic curves in projective space, together with an interpretation vis-a-vis the Wirtinger theorem, is given in [14].

The classical order functions are extensively discussed in Nevanlinna [32], to which we also refer for manifold uses of the Poisson-Jensen-Nevanlinna formula. Theorem 1.27 is due to Hadamard; a very elegant proof using the Poisson-Jensen-Nevanlinna formula is in [32], where more detailed properties of Weierstrass primary factors are presented. The n-variable version of (1.27) is due to Lelong [30] and Stoll [39]; our argument appears to be similar to that of Stoll.

The problem of growth properties of entire analytic sets of higher co-dimension is in an unclear state. On the one hand there is the Bezout counterexample of Cornalba-Shiffman [15], and on the other the existence theorems of Pan and Skoda [36]. Further discussion is given in [20], in the book [40] by Stoll, and in Carlson [7].

CHAPTER 2
THE APPEARANCE OF CURVATURE

(a) *Heuristic reasoning*

In the first chapter we discussed various ways of measuring the growth of entire analytic sets and entire holomorphic mappings. The First Main Theorem provides the basic relation between growth of $f : C^n \to M$ and the growth of divisors $D_f = f^{-1}(D)$. For an entire meromorphic function $f(z)$, a consequence of this result is that the number $n_f(a, r)$ of solutions to the equation $f(z) = a$ in the disc $|z| \leq r$ can grow no faster than the average number of solutions to this equation. In general, because of the somewhat intractible character of the proximity form $m_f(D, r)$, the F.M.T. has thus far been only used to estimate the counting function $N_f(D, r)$ from above by the order function $T_f(L, r)$ $(D \in |L|)$, and not to discuss the deeper Picard-type question of how large a divisor $f(C^n)$ can miss.

An exception was E. Borel's proof of the little Picard theorem given in Chapter 1. For this argument we not only needed a good formalism for measuring the growth of entire meromorphic functions, but also required information on the growth of the derivative of an entire *holomorphic* function. In order to discuss the more profound question of obtaining a lower bound on

$$n_f(a, r) + n_f(b, r) + n_f(c, r)$$

$(a, b, c$ being three distinct points on the sphere), one needs to bound the growth of $f'(z)$ for an entire *meromorphic* function $f(z)$. A related problem is to estimate the number of ramification points of $f : C \to P^1$ in terms of the growth of f. Questions of this sort lead naturally into curvature considerations, as we hope to now make evident.

In a somewhat general fashion, we consider a holomorphic mapping

$$f : C \to S$$

into a compact Riemann surface on which one is given a conformal metric ds^2. In terms of a local holomorphic coordinate w on S

$$ds^2 = h(w)|dw|^2 \; ,$$

and the Gaussian curvature is

$$K(w) = \frac{-4}{h(w)} \frac{\partial^2 \log h(w)}{\partial w \, \partial \overline{w}} \; .$$

Writing

$$ds_f^2 = f^*(ds^2) = k(z)|dz|^2$$

where

$$k(z) = |w'(z)|^2 \, h(w(z)) \; ,$$

we see that ds_f^2 is a pseudo-metric on C having isolated zeroes on the ramification divisor R_f of f and Gaussian curvature $K(z) = K(f(z))$. Denoting by $dA_f = \frac{\sqrt{-1}}{2\pi} k \, dz \wedge d\overline{z}$ the pullback to C of the area element on S,

$$dd^c \log k = -K dA_f \; .$$

Applying the proof of the F.M.T. (1.15) to the function $k(z)$ then gives

$$(2.1) \qquad N(R_f, r) = \int_0^r \left(\int_{\Delta(\rho)} K dA_f \right) \frac{d\rho}{\rho} + \frac{1}{2\pi} \int_0^{2\pi} \log k(re^{i\theta}) \, d\theta + C \; .$$

This is the Second Main Theorem (S.M.T.) in the present context, and formulae of this general type will be systematically used to give inequalities going in the opposite direction to (1.16).

To explain how (2.1) may be used to estimate $N(R_f, r)$, we assume the upper bound

$$K \leq K_0$$

on the Gaussian curvature and denote by

$$T_f(r) = \int_0^r \left(\int_{\Delta(\rho)} dA_f \right) \frac{d\rho}{\rho}$$

the Ahlfors-Shimizu order function for the metric ds^2. The growth of $T_f(r)$ is determined up to the multiplicative constant $\int_S dA$, for the same reason that the growth of $T_f(L, r)$ was seen to be intrinsic in Chapter 1. We shall prove that, given $\varepsilon > 0$,

$$(2.2) \qquad\qquad N(R_f, r) \leqq (K_0 + \varepsilon) T_f(r) \qquad //_\varepsilon \ ,$$

where the notation $//_\varepsilon$ means that

"the stated inequality holds outside an open

set $I_\varepsilon \subset R^+$ with $\displaystyle\int_{I_\varepsilon} \frac{dt}{t} < \infty$."

The derivation of (2.2) from (2.1) proceeds by two steps which are ubiquitous in Nevanlinna theory.

The first is the *concavity of the logarithm*:

$$(2.3) \qquad \frac{1}{2\pi} \int_0^{2\pi} \log u(\theta) \, d\theta \leqq \log \left(\frac{1}{2\pi} \int_0^{2\pi} u(\theta) \, d\theta \right)$$

for a function $u(\theta) \geqq 0$ such that $\log u$ is integrable. This is proved from the finite form:

$$(u_1 \cdots u_n)^{\frac{1}{n}} \leqq \frac{1}{n} (u_1 + \cdots + u_n)$$

$$\Downarrow$$

$$\frac{1}{n} (\log u_1 + \cdots + \log u_n) \leqq \log \left(\frac{1}{n} (u_1 + \cdots + u_n) \right)$$

by passing to the limit. Applying (2.3),

$$\frac{1}{2\pi}\int_0^{2\pi} \log k(re^{i\theta})\,d\theta \leq \log\left(\frac{1}{2\pi}\int_0^{2\pi} k(re^{i\theta})\,d\theta\right) = \log\left[\frac{1}{r^2}\frac{d^2 T_f(r)}{(d \log r)^2}\right],$$

so that from the S.M.T. (2.1)

(2.4) $$N(R_f, r) \leq K_0 T_f(r) + \log\left[\frac{1}{r^2}\frac{d^2 T_f(r)}{(d \log r)^2}\right].$$

If $T_f(r)$ were a nice growth function, such as $\log^\alpha r \cdot r^\beta \cdot e^{r^\gamma}$, then

$\dfrac{1}{r^2}\dfrac{d^2 T_f(r)}{d \log r^2}$ is evidently $0(T_f(r))$, and this proves (2.2) (without ex-

ceptional intervals). In general, we shall use the case $n = 1$ and $S(r) = T_f(r)$ from the following *calculus lemma*:

(2.5) LEMMA. *Let* $S(r)$ *be a positive function such that* $r^{2n-1}\dfrac{dS(r)}{dr}$
is positive and increasing and $\dfrac{d^2 S(r)}{dr^2}$ *is continuous. Then, given positive*
ε *and* δ,

$$r^{2n-1}\frac{d}{dr}\left(r^{2n-1}\frac{dS(r)}{dr}\right) \leq r^{4n-2+\varepsilon}(S(r))^{1+\delta} \qquad //_{\varepsilon,\delta}.$$

PROOF. We shall use twice the assertion: Let $f(r)$, $g(r)$, $a(r)$ be positive increasing functions of r such that $g'(r)$ is continuous, $f'(r)$ is piecewise continuous, and $\int^\infty \dfrac{dr}{a(r)} < \infty$. Then

(2.6) $$f'(r) \leq g'(r)a(f(r))$$

outside an open set $I \subset R^+$ such that

$$\int_I dg \leq \int^\infty \frac{dr}{a(r)}.$$

$\Bigg($ *Proof.* On the set I where (2.6) fails to hold, $g'(r) < \dfrac{f'(r)}{a(f(r))}$ and consequently

$$\int_I dg < \int_I \frac{f'(r)\,dr}{a(f(r))} \leq \int^\infty \frac{f'(r)\,dr}{a(f(r))} = \int^\infty \frac{dr}{a(r)}.\Bigg)$$

Applying (2.6) with $f(r) = S(r)$, $g(r) = \frac{r^\mu}{\mu}$, $a(r) = r^\nu$ and $\mu, \nu > 1$ gives

$$S'(r) \leq r^{\mu-1} (S(r))^\nu \qquad //\,.$$

Keeping the same g and a and setting $f = r^{2n-1}S'(r)$ gives

$$\frac{d}{dr}\left(r^{2n-1} \frac{dS(r)}{dr}\right) \leq r^{\mu-1} r^{(2n-1)\nu}(S'(r))^\nu \qquad //\,.$$

Combining these and readjusting constants yields

$$r^{2n-1} \frac{d}{dr}\left(r^{2n-1} \frac{dS(r)}{dr}\right) \leq r^{4n-2+\varepsilon} (S(r))^{1+\delta} \qquad //_{\varepsilon,\delta}\,,$$

as was to be proved.

Proof of (2.2). By the calculus lemma,

$$\log\left[\frac{1}{r^2} \frac{d^2 T_f(r)}{(d \log r)^2}\right] \leq \varepsilon \log r + (1+\varepsilon) \log T_f(r) \qquad //_\varepsilon\,.$$

On the other hand, for large r $\quad T_f(r) \geq C \log r \ (C > 0)$, and the result follows from this together with (2.4).

In case $S = \mathbf{P}^1$ is the Riemann sphere with coordinate w, the standard metric

$$ds^2 = \frac{\sqrt{-1}}{2\pi} \frac{|dw|^2}{(1+|w|^2)^2}$$

has constant Gaussian curvature $+2$, and so

(2.7) $$\lim_{r \to \infty} \frac{N(R_f, r)}{T_f(r)} \leq 2\,.$$

The Weierstrass $\wp$-function associated to a lattice $\Lambda = \{m\omega_1 + n\omega_2 : m, n \in \mathbf{Z}\}$ in the plane is a doubly-periodic entire meromorphic function. In each fundamental parallelogram, $\wp(z)$ is a 2-to-1 mapping onto $\mathbf{P}^1$ and has 4 points of ramification. Thus

$$N(R_f, r) = 2T_f(r) + o(T_f(r))$$

so that (2.7) is sharp.

We next wish to illustrate how negative curvature may be used. Suppose that S is a compact Riemann surface of genus $g(S) > 1$, which is the same as saying that the canonical line bundle K_S is positive. Then, by linear potential-theoretic methods (cf. Proposition 2.17 below) we may find a metric on S whose Gaussian curvature satisfies $K \leq -1$. The S.M.T. (2.1) then gives

$$T_f(r) = \int_0^r \left(\int_{\Delta(\rho)} dA_f \right) \frac{d\rho}{\rho} \leq \int_0^r \left(\int_{\Delta(\rho)} -KdA_f \right) \frac{d\rho}{\rho} \leq C \log T_f(r) \quad //$$

under the assumption that f is non-constant. Since $T_f(r) \to \infty$, this is clearly impossible, and we have proved the following result also due to Picard:

A holomorphic mapping $f : C \to S$ *with* $g(S) \geq 2$ *is necessarily constant.*

Of course this theorem is trivial if one assumes the uniformization theorem, which is equivalent to finding a metric of constant Gaussian curvature $K \equiv -1$. This latter is, however, a deep theorem which basically involves solving a non-linear equation of the type

$$\Delta u = e^u .$$

Moreover, the method will not work in several variables. Finding a metric with $K \leq -1$ boils down to solving a linear equation of the type (cf. Lemma 1.13)

$$\Delta u = v$$

on S, and this will generalize to compact Kähler manifolds.

At this stage, however, we mainly wish to point out that curvature appears quite naturally in the theory, and that the presence of a negatively curved metric has very strong function-theoretic implications.

(b) *Volume forms*

On a complex manifold M of dimension n, a *volume form* Ψ is a smooth positive (n, n) form. In a local holomorphic coordinate system $z = (z_1, \cdots, z_n)$,

$$\Psi(z) = h(z)\Phi(z)$$

where

$$\Phi(z) = \prod_{j=1}^{n} \frac{\sqrt{-1}}{2\pi} (dz_j \wedge d\bar{z}_j)$$

is the Euclidean volume form and h(z) is a positive C^∞ function. A *pseudo-volume form* is a smooth non-negative (n,n) form Ψ having a similar local expression, but whose coefficient function is

$$h(z) = |g(z)|^2 k(z)$$

with $g(z) \not\equiv 0$ being a holomorphic function and k(z) being positive. Pseudo-volume forms arise by pulling back volume forms under equi-dimensional, non-degenerate holomorphic mappings f. The *ramification divisor* R_f of such a mapping is given locally by $g(z) = 0$.

The *Ricci form* of a pseudo-volume form Ψ is the smooth (1,1) form given locally by

$$\text{Ric } \Psi = dd^c \log k(z) \,.$$

Since

$$dd^c \log |j(z)|^2 = 0$$

for a non-zero holomorphic function j(z), it follows that $\text{Ric } \Psi$ is globally defined on M and that

$$\text{Ric} (f^* \Psi) = f^*(\text{Ric } \Psi)$$

for an equi-dimensional, non-degenerate holomorphic mapping f. Use of the formula

(2.8) $$\text{Ric} (e^u \Psi) = \text{Ric } \Psi + dd^c u$$

affords easy manipulation of Ricci forms. In particular,

(2.9) $\text{Ric}(C\Psi) = \Psi$ $(C > 0 \text{ a constant})$,

so that we may always adjust constants.

Volume forms which satisfy the curvature estimates

(2.10)
$$\begin{cases} \text{Ric } \Psi \geqq 0 \\ (\text{Ric } \Psi)^n \geqq \Psi \end{cases}$$

play a crucial role in the theory. These inequalities are evidently in-variant under non-degenerate holomorphic mappings. Here are some examples.

   (i)   When $n = 1$ so that M is a Riemann surface, there is a natural 1-1 correspondence between conformal metrics and volume forms given by

$$ds^2 = h(z)|dz|^2 \leftrightarrow \frac{\sqrt{-1}}{2\pi} h(z)\,dz \wedge d\bar{z} = \Psi\ .$$

For such a Ψ,
$$\text{Ric } \Psi = (-K)\Psi$$

where K is the Gaussian curvature of ds^2. Hence, in this case (2.10) says that ds^2 has Gaussian curvature $K \leqq -1$.

   In particular, on the disc $\Delta(R) = \{z \in C : |z| < R\}$, the *Poincaré metric*

(2.11) $$\Theta(R) = \frac{\sqrt{-1}}{\pi}\frac{R^2\,dz \wedge d\bar{z}}{(R^2 - |z|^2)^2}$$

satisfies

(2.12) $$\text{Ric } \Theta(R) = \Theta(R)\ .$$

   We write $\Delta = \Delta(1)$ and $\Theta = \Theta(1)$. The punctured disc $\Delta^* 9 \{\zeta \in C : 0 < |\zeta| < 1\}$ has Δ as universal covering, and the Poincaré metric there induces

(2.13) $$\Theta^* = \frac{\sqrt{-1}}{\pi}\frac{d\zeta \wedge d\bar{\zeta}}{|\zeta|^2 (\log|\zeta|^2)^2}$$

on Δ^* such that (2.12) is satisfied for Θ^*.

(ii) On the polycylinder

$$\begin{cases} \Delta^n(R) = \{(z_1, \cdots, z_n) \, \epsilon \, \mathbb{C}^n : |z_j| < R_j\} \\ R = (R_1, \cdots, R_n) \text{ with } R_j > 0 \,, \end{cases}$$

the product of Poincaré metrics induces a volume form $\Theta_n(R)$ such that

(2.14)
$$\begin{cases} \text{Ric } \Theta_n(R) > 0 \\ \{\text{Ric } \Theta_n(R)\}^n = \Theta_n(R) \,. \end{cases}$$

As above, $\Theta_n = \Theta_n(1, \cdots, 1)$ on the unit polycylinder induces a volume form satisfying (2.14) on the punctured polycylinders

$$\Delta^*_{k,n} = (\Delta^*)^k \times \Delta^{n-k} \,.$$

(iii) On a general complex manifold M, a volume Ψ is the same as a metric for the dual K^*_M of the canonical line bundle, and $\text{Ric } \Psi$ is the Chern form of K_M for this metric. When M is compact, we may find a volume form satisfying the curvature conditions (2.10) exactly when K_M is positive in the sense of Kodaira. (*Proof*: If ω is a positive (1.1) form representing $c_1(K_M)$ in $H^2(M, \mathbb{R})$, then by Lemma 1.13 we may find a metric in K_M whose Chern form is ω. Letting Ψ' be the corresponding volume form, $\text{Ric } \Psi' = \omega$ and so $(\text{Ric } \Psi')^n \geq C\Psi'$ $(C > 0)$ by the compactness of M. Now adjust constants to obtain (2.10).)

This suggests that on a general compact, complex manifold M, we should look for such "negatively curved" volume forms on Zariski open sets

$$M - D$$

where D is an effective divisor such that $K_M \otimes [D]$ is positive. Some restrictions on the singularities of D are necessary, and so we assume that D has *simple normal crossings* in the sense that

$$D = D_1 + \cdots + D_N$$

where the D_j are smooth divisors meeting transversely. In a neighborhood $\mathcal{U}$ of a point $p \in M$ through which exactly k of the divisors pass, we may choose local holomorphic coordinates $(z_1, \cdots, z_n)$ such that $D \cap \mathcal{U}$ is given by the equation

$$z_1 \cdots z_k = 0 .$$

Thus $(M-D) \cap \mathcal{U} = \Delta^*_{k,n}$ is a punctured polycylinder, and this suggests the following global version of the Poincaré volume form on $\Delta^*_{k,n}$ (cf. (2.13)): Choose a volume form Ψ_M on M and metrics in the line bundles $[D_j]$ such that the inequality of Chern forms

(2.15) $$c_1(K_M) + c_1([D]) > 0$$

is satisfied (recall that $[D] \cong \bigotimes\limits_{j=1}^{N} [D_j]$ and

$$c_1([D]) = c_1([D_1]) + \cdots + c_1([D_N])) .$$

Let $s_j \in \mathcal{O}(M, [D_j])$ be a holomorphic section which defines D_j, and set

(2.16) $$\Psi_M(D) = \frac{\Psi_M}{\prod\limits_{j=1}^{N} |s_j|^2 (\log(\alpha|s_j|^2))^2} .$$

(2.17) PROPOSITION. *Given* $\varepsilon > 0$, *there exists* $\beta > 0$ *such that for* $0 < \alpha \le \beta$ *the volume form* $\Psi_M(D)$ *is* C^∞ *on* $M-D$ *and satisfies*

$$\begin{cases} \text{Ric } \Psi_M(D) = (1-\varepsilon)\{c_1(K_M) + c_1([D])\} + \psi , \\ \psi > 0 \text{ and } \psi^n \ge \Psi_M(D) . \end{cases}$$

In particular, the curvature conditions (2.10) *are satisfied.*

REMARK. The reason we have replaced (2.10) by the excess relation above is to remove the ambiguity caused by the possibility of adjusting constants according to (2.9).

PROOF. Set $\omega = c_1(K_M) + c_1([D]) > 0$. Then

$$
\begin{cases}
\text{Ric } \Psi_M = c_1(K_M) \\[2ex]
dd^c \log \dfrac{1}{a|s_j|^2} = c_1([D_j]) \\[2ex]
\displaystyle\sum_{j=1}^{N} c_1([D_j]) = c_1([D]) .
\end{cases}
$$

Using these relations in (2.8) gives

$$
\text{Ric } \Psi_M(D) = \omega - \sum_{j=1}^{N} dd^c \log \left(\log a|s_j|^2\right)^2
$$

$$
= (1 - 2\varepsilon)\omega + \psi
$$

where

$$
\psi = 2\varepsilon \cdot \omega - \sum_{j} dd^c \log \left(\log a|s_j|^2\right)^2 .
$$

Now then

$$
dd^c \log \left(\log a|s_j|^2\right)^2 = \frac{-2 dd^c \log a|s_j|^2}{\left(\log a|s_j|^2\right)} + \frac{4 d \log a|s_j|^2 \wedge d^c \log a|s_j|^2}{\left(\log a|s_j|^2\right)^2} .
$$

The first term on the right is

$$
\frac{2 c_1([D_j])}{\log a|s_j|^2}
$$

which tends to zero as $a \to 0$ and may thus be absorbed in $\varepsilon\omega$. Thus, for a sufficiently small

$$
\psi \geq \varepsilon\omega + \sum_{j=1}^{N} \frac{\sqrt{-1}}{\pi} \frac{\partial \log (a|s_j|^2) \wedge \bar{\partial} \log (a|s_j|^2)}{\left(\log a|s_j|^2\right)^2} > 0
$$

since each term in the sum is non-negative.

We now localize around a point $p \in D$. We may choose a coordinate neighborhood $\mathcal{U}$ of p in which D has the equation $z_1 \cdots z_k = 0$. In fact, we may assume that $D_j \cap \mathcal{U}$ is given by $z_j = 0$ for $j = 1, \cdots, k$, and that $D_{k+1}, \cdots, D_N$ do not meet $\mathcal{U}$. Then

$$(2.18) \qquad \psi \geq \varepsilon \cdot \omega + \sum_{j=1}^{k} \frac{\sqrt{-1}}{\pi} \frac{\partial \log (a|s_j|^2) \wedge \bar{\partial} \log (a|s_j|^2)}{(\log a|s_j|^2)^2} .$$

Now $a|s_j|^2 = a\gamma_j(z)|z_j|^2$ where γ_j is a positive C^∞ function. Consequently

$$(2.19) \qquad \partial \log (a|s_j|^2) \wedge \bar{\partial} \log (a|s_j|^2) = \frac{dz_j \wedge d\bar{z}_j}{|z_j|^2} + \rho_j$$

where

$$(2.20) \qquad \rho_j = \frac{\partial \gamma_j \wedge \bar{\partial}\gamma_j}{\gamma_j^2} + \frac{\partial \gamma_j \wedge d\bar{z}_j}{\bar{z}_j \gamma_j} + \frac{dz_j \wedge \bar{\partial}\gamma_j}{z_j \gamma_j}$$

has the property that $|z_j|^2 \rho_j$ is C^∞ and vanishes on D_j. Since

$$\omega \geq \frac{C\sqrt{-1}}{2\pi} \left(\sum_{j=1}^{n} dz_j \wedge d\bar{z}_j \right) \quad \text{for some } C > 0, \text{ it follows from } (2.18)-(2.20)$$

$$(\psi(z))^n \geq C' \left\{ \frac{\Phi(z) + \Lambda(z)}{\prod_j |z_j|^2 (\log a|s_j|^2)^2} \right\}$$

where $\Phi(z)$ is the Euclidean volume form and $\Lambda(z)$ is C^∞ and vanishing on D. By (2.16)

$$\psi^n \geq C'' \Psi_M(D)$$

holds in a perhaps smaller neighborhood $\mathcal{U}'$ of p. By compactness, we may assume that $\psi^n \geq C''' \Psi_M(D)$ $(C''' > 0)$ on all of M, and then we may make $C''' = 1$ by adjusting constants.   Q.E.D.

As a special case, we consider complex projective space $\mathbf{P}^n$ with homogeneous coordinates $Z = [\mathfrak{z}_0, \cdots, \mathfrak{z}_n]$ and standard Kähler form

$$\omega = dd^c \log \|Z\|^2 .$$

Then $\omega^n = \Psi$ is a volume form and

$$\text{Ric } \Psi = -(n+1)\omega .$$

$\Big($Proof: In affine coordinates $Z = [1, w_1, \cdots, w_n]$,

$$\omega = \frac{\sqrt{-1}}{2\pi} \partial\bar{\partial} \log (1 + \|w\|^2)$$

$$= \frac{\sqrt{-1}}{2\pi} \frac{(dw, dw)}{1 + \|w\|^2} - \frac{\sqrt{-1}}{2\pi} \frac{(dw, w)(w, dw)}{(1 + \|w\|^2)^2}$$

$$= \alpha + \beta$$

where α is the first term and β the second. Since $\beta \wedge \beta = 0$,

$$\omega^n = \alpha^n + n\alpha^{n-1} \vee \beta$$

$$= \left\{ n! - \frac{n(n-1)! \, \|w\|^2}{1 + \|w\|^2} \right\} \frac{\Phi(w)}{(1 + \|w\|^2)^n}$$

$$= \frac{n! \, \Phi(w)}{(1 + \|w\|^2)^{n+1}} .$$

It follows that

$$\text{Ric } \Psi = -(n+1)dd^c \log (1 + \|w\|^2) = -(n+1)\omega . \Big)$$

We assume that the divisors D_j are linear hyperplanes given by equations

$$<A_j, Z> = 0 .$$

The condition that $D = D_1 + \cdots + D_N$ have simple normal crossings is that these hyperplanes should be in general position. Since

$$dd^c \log \left(\frac{\|A_j\|^2 \|Z\|^2}{|<A_j, Z>|^2} \right) = \omega ,$$

the Chern form $c_1([D]) = N \cdot \omega$ and the positivity condition (2.15) is

$$N > n+1 .$$

In summary: *On* $P^n - \{N \geq n+2$ *hyperplanes in general position*$\}$, *there is a volume form which satisfies* (2.10). *In particular, this is true on* $P^1 - \{N \geq 3$ *distinct points*$\}$.

As another example, if M is a compact Riemann surface of genus g with distinct marked points $\{x_1, \cdots, x_N\}$, then (2.16) gives a metric ds^2 on

$$M - \{x_1, \cdots, x_N\} \qquad (N+2g-2 > 0)$$

having Gaussian curvature $K \leq -1$.

We may use this metric Ψ on $P^1 - \{a, b, c\}$, together with the S.M.T. (2.1), to give another proof of the Picard theorem as follows: Using Proposition (2.17), write

$$\begin{cases} \text{Ric } \Psi = (1-\varepsilon)\omega + \psi & \text{where} \\ \\ \psi \geq \Psi \end{cases}$$

and ω is the standard metric on P^1. If a non-constant holomorphic mapping $f : C \to P^1 - \{a, b, c\}$ exists, then we set

$$S(r) = \int_0^r \left(\int_{\Delta(\rho)} f^*\psi \right) \frac{d\rho}{\rho}$$

and apply (2.1) just as at the end of (a) in Chapter 2 to obtain an inequality

$$(1-\varepsilon)T_f(r) + S(r) \leq \log (S(r)) \qquad // ,$$

which is a contradiction.

(c) *The Ahlfors lemma*

Among the most subtle and far-reaching tools in the study of holomorphic mappings is Ahlfors' generalization of the Schwarz lemma:

(2.21) AHLFORS LEMMA. *Let* Ψ *be a pseudo-volume form on the poly-cylinder* $\Delta^n(R)$ *which satisfies the curvature estimate* (2.10). *Then*

$$\Psi \leq \Theta_n(R) .$$

PROOF. For $r \leq R$ we write

$$\Psi(z) = u_r(z)\Theta_n(r) \qquad (z \in \Delta^n(r)) .$$

It will suffice to prove that $u_r \leq 1$ for $r < R$, because $\lim\limits_{r \to R} u_r(z) = u_R(z)$ for fixed $z \in \Delta^n(R)$. For $r < R$, $u_r(z)$ tends to zero as $z \to \partial\Delta^n(r)$ since $\Theta_n(r)(z)$ goes to infinity there. Thus $u_r(z)$ has an interior maximum at some point z_0, and by (2.8) and the maximum principle

$$0 \geq \frac{\sqrt{-1}}{2\pi} \partial\bar\partial \log u_r(z_0) = \text{Ric } \Psi(z_0) - \text{Ric } \Theta(r)(z_0) .$$

Using (2.10) and (2.14) and taking determinants in this inequality gives

$$\Psi(z_0) \leq \text{Ric } \Psi(z_0)^n \leq \text{Ric } \Theta(r)(z_0)^n = \Theta(r)(z_0) ,$$

which exactly says that $u_r(z_0) \leq 1$.   Q.E.D.

For a holomorphic mapping

$$f : \Delta \to \Delta$$

we may apply this lemma to the pseudo-metric $f^*(\Theta)$ and obtain Pick's invariant form of the Schwarz lemma

$$\frac{|f'(z)|}{1 - |f(z)|^2} \leq \frac{1}{1 - |z|^2} .$$

In case $f(0) = 0$ this reduces to

$$|f'(0)| \leq 1 ,$$

which is the usual Schwarz lemma.

An application of the Ahlfors lemma is the

(2.22) SCHOTTKY-LANDAU THEOREM. *Let* $f : \Delta^n(R) \to M$ *be a non-degenerate, equidimensional holomorphic mapping into a complex manifold* M *having a volume form* Ψ *satisfying (2.10). Writing* $f^*\Psi(z) = J_f(z)\Phi(z)$ *where* Φ *is the Euclidean volume,*

$$R_1 \cdots R_n \leq C J_f(0)^{-\frac{1}{2}} \qquad R = (R_1, \cdots, R_n) .$$

PROOF. By (2.11) and the Ahlfors lemma at $z = 0$

$$J_f(0) \leq \left(\frac{1}{2}\right)^n \frac{1}{R_1^2 \cdots R_n^2} .$$

Applying Proposition 2.7 gives the

(2.23) COROLLARY. *Let* M *be a compact, complex manifold and* D *a divisor with simple normal crossings such that*

$$c_1(K_M) + c_1([D]) > 0 .$$

Then any non-degenerate, equidimensional holomorphic mapping

$$f : \Delta^n(R) \to M - D$$

with non-zero Jacobian at $z = 0$ *satisfies*

$$R_1 \cdots R_n \leq C ,$$

where the constant C *depends only on the magnitude of the Jacobian of* f *at* $z = 0$. *In particular, an entire holomorphic mapping*

$$f : C^n \to M - D$$

is necessarily degenerate.

Taking $M = P^n$ and D to be a set of $n+2$ hyperplanes A_j in general position, we find the theorem of A. Bloch, reproved independently by Mark Green, that an entire holomorphic mapping

$$f : C^n \to P^n - \{A_1 + \cdots + A_{n+2}\}$$

is degenerate. For $n = 1$ this is the Picard theorem. The Schottky-Landau theorem gives both results in finite form.

It is perhaps of interest to compare the Ahlfors lemma with the methods based on the integral formula (2.1). Both imply that if $ds^2 = h(z)|dz|^2$ is a metric on the disc $\Delta(r)$ whose Gaussian curvature satisfies $K \leqq -1$, then $r < \infty$. From the Ahlfors lemma, one finds the bound

$$(2.24) \qquad\qquad r \leqq Ch(0)^{-\frac{1}{2}} .$$

On the other hand, setting

$$S(r) = \int_0^r \left\{ \int_{\Delta(\rho)} h(z) \frac{\sqrt{-1}}{2\pi} (dz \wedge d\bar{z}) \right\} \frac{d\rho}{\rho} ,$$

the S.M.T. (2.1) gives

$$S(r) \leqq \log\left[\frac{1}{r^2} \frac{d^2 S(r)}{(d \log r)^2}\right] .$$

Then the calculus lemma (2.5) implies that $r \neq +\infty$. Being more careful in the proof of this lemma, one may also prove (2.24) by this method. The advantage of integral formulae is that this technique also works for singular volume forms obtained by pulling back (2.16) under a non-degenerate, equidimensional holomorphic mapping

$$f : C^n \to M ,$$

and leads to a lower bound on the size of $D_f = f^{-1}(D)$.

Actually, the Ahlfors lemma implies, at least in principle, such a lower bound on the size of D_f as follows: First, for a point $z \in C^n$ we let

$B[z, r]$ be the ball of radius r centered at z. If $V \subset C^n$ is a k-dimensional analytic set, then (1.6) gives the estimate

$$(2.25) \qquad \text{vol}(V \cap B[z, r]) \geq cr^{2k}$$

for any point $z \in V$. Next, for an entire holomorphic mapping $f : C^n \to M$ as above, we set

$$f^* \Psi_M(D) = J_f \Phi$$

where $\Psi_M(D)$ is the singular volume form (2.16) and Φ is the Euclidean volume form on C^n. If $\Delta_n(z, R)$ is the polycylinder $|w_j - z_j| < R_j$, then for any such $\Delta_n(z, R)$ contained entirely in $C^n - D_f$, (2.23) gives

$$R_1 \cdots R_n \leq C_{J_f}(z)^{-\frac{1}{2}} .$$

Thus, for any point $z \in C^n - D_f$ there is a bound on what size and shape polycylinder centered at z may be put in $C^n - D_f$. This bound depends only on the size of $J_f(z)$, and so if $J_f(z)$ bounded from below off a relatively small subset of C^n, we obtain a lower bound on the size of D_f using (2.25). By the remarks in Section (a) of this chapter, we may expect that $J_f(z)$ is small only on a subset of C^n which is small relative to the growth of f, thus putting a lower bound on the size of D_f.

It seems difficult to make this heuristic reasoning precise, and so we shall rely on the more analytical integral formulae.

Before going to the general Second Main Theorem, we shall combine the Ahlfors lemma with the integral formula method to prove the

BIG PICARD THEOREM. *A holomorphic mapping*

$$f : \Delta^* \to P^1 - \{a, b, c\}$$

extends across the origin to a holomorphic mapping into P^1.

PROOF. We let $\Delta^*(r)$ be the punctured disc $\{0 < |z| < r\}$ and $A(r)$ the annulus $\frac{1}{r} \leq |z| \leq 1$. By change of scale, we may assume that f is defined on $\Delta^*(1+\varepsilon)$ for some $\varepsilon > 0$.

Let $u(z)$ be a non-negative C^∞ function on $\Delta^*(1+\varepsilon)$ having the local form

$$\begin{cases} u(z) = |z-z_0|^{-2\mu(z_0)} u_{z_0}(z) \\[2mm] u_{z_0} > 0 \quad \text{and} \quad \mu(z_0) \geq 0 \end{cases}$$

around any point z_0. We let D be the divisor

$$\sum \mu(z_0) \cdot z_0$$

and $n(D,r)$ the counting function $\sum_{z_0 \epsilon A(r)} \mu(z_0)$. Then the proof of (1.15) gives

$$(2.26) \quad N(D,r) + \frac{1}{2\pi} \int_{|z|=\frac{1}{r}} \log u \cdot d\theta = \int_1^r \left(\int_{A(\rho)} dd^c \log u \right) \frac{d\rho}{\rho} + 0(\log r)$$

where $N(D,r) = \int_1^r n(D,\rho) \frac{d\rho}{\rho}$.

Let $A \epsilon P^1$ be a point and $A_f = f^{-1}(A)$ the inverse image. Then the estimate

$$n(A_f, r) = 0(1) \iff N(A_f, r) = 0(\log r)$$

implies that A_f is finite; and if this is true for any $A \epsilon P^1$ the Casorati-Weierstrass theorem implies that f extends as desired. To derive this estimate on $N(A_f, r)$, we use homogeneous coordinates and write

$$\begin{cases} f(z) = Z(z) = [z_0(z), z_1(z)] \\[2mm] A = [a_0, a_1] \\[2mm] u(z) = \dfrac{\|Z(z)\|^2 \|A\|^2}{|<A, Z(z)>|^2} \geq 1 \ . \end{cases}$$

Then $dd^c \log u(z) = f^*\omega = \omega_f$ is the pull-back of the standard Kähler form on P^1, and (2.26) gives

(2.27) $$N(A_f, r) \leq \int_0^r \left(\int_{A(\rho)} \omega_f \right) \frac{d\rho}{\rho} + 0(\log r) .$$

Now then by inspection

$$\omega \leq C\Psi_{\mathbf{P}^1}(\{a, b, c\})$$

for the singular volume form (2.16) in this case. Moreover,

$$\Psi_{\mathbf{P}^1}(\{a, b, c\}) \leq \frac{\sqrt{-1}}{2\pi} \frac{dz \wedge d\bar{z}}{|z|^2 (\log(1+\varepsilon)|z|^2)^2}$$

by the Ahlfors lemma. Since this Poincaré volume form on $\Delta^*(1+\varepsilon)$ has finite integral over $\Delta^*(1)$, it follows that

$$\int_{\Delta^*(1)} \omega_f = 0(1) \Longrightarrow \int_1^r \left(\int_{A(\rho)} \omega_f \right) \frac{d\rho}{\rho} = 0(\log r) ,$$

which when combined with (2.27) proves our result.

REMARK. The same proof works for an equi-dimensional, non-degenerate holomorphic map

$$f : \Delta^*_{k,n} \to M - D$$

where M and D satisfy the hypotheses in (2.23), and gives a meromorphic extension of f to Δ^n.

(d) *The Second Main Theorem*

On $\mathbf{C}^n$ we consider a *singular volume form*

$$\Lambda(z) = \lambda(z)\Phi(z)$$

where $\Phi(z)$ is the Euclidean volume and where the coefficient function has the local form

(2.28) $$\lambda(z) = \frac{|g(z)|^2 h(z)}{\displaystyle\prod_{j=1}^{k} |f_j(z)|^2 \log(a_j(z)|f_j(z)|^2)^2}$$

with $g(z)$, $f_j(z)$ being non-identically zero holomorphic functions and $h(z)$, $a_j(z)$ being positive C^∞ functions. We also assume that $\lambda(z)$ is C^∞ around the origin. Of course, we have in mind Λ's of the type $f^*(\Psi)$ where $f : C^n \to M$ is a non-degenerate, equi-dimensional holomorphic mapping and Ψ is given by (2.16). The *ramification divisor* R and *singular divisor* D are respectively defined by

$$\begin{cases} g(z) = 0 \ , \\ f(z) = 0 \quad \text{where} \quad f(z) = \prod_{j=1}^{k} f_j(z) \ . \end{cases}$$

The general Second Main Theorem (S.M.T.) is

$$(2.29) \quad \int_0^r \left\{ \int_{B[\rho]} \text{Ric } \Lambda \wedge \omega^{n-1} \right\} \frac{d\rho}{\rho} + N(R, r) = N(D, r) + \int_{\partial B[r]} \log \lambda \cdot \Sigma + C \ .$$

PROOF. By averaging over the lines through the origin as in the proof of (1.5), it will suffice to prove the case $n = 1$. Were it not for the $(\log a_j |f_j|^2)^{-2}$ terms, (2.29) would be the same as (1.5), with the trivial modification that Jensen's theorem for meromorphic functions must be utilized. What must therefore be proved is

$$\int_0^r \left\{ \int_{\Delta(\rho)} dd^c \log (\log a|f|^2) \right\} \frac{d\rho}{\rho} = \frac{1}{2\pi} \int_{|z| = \rho} \log (\log a|f|^2)^2 d\theta + C \ ,$$

where f is holomorphic and a is positive and C^∞. Setting $\beta = a|f|^2$, the equations

$$\bar{\partial} \log (\log \beta) = \frac{\partial \log a}{\log \beta} + \frac{d\bar{f}}{\bar{f} \log \beta}$$

$$\partial\bar{\partial} \log (\log \beta) = \frac{\partial\bar{\partial} \log \beta}{\log \beta} - \frac{\partial \log \beta \wedge \bar{\partial} \log \beta}{(\log \beta)^2}$$

show first that $dd^c \log (\log a|f|^2)^2 = \frac{\sqrt{-1}}{2\pi} \partial\bar{\partial} \log (\log \beta)$ is integrable.

Next, Stokes' theorem may be applied to $d(\bar{\partial} \log (\log \beta)) = \partial\bar{\partial} \log (\log \beta)$ by cutting out ε-discs around the zeroes of f, then applying Stokes' theorem to the complement of these discs, and finally observing that

$$\lim_{\varepsilon \to 0} \int_{|z|=\varepsilon} \frac{d\bar{z}}{\log |z|} = 0 = \lim_{\varepsilon \to 0} \int_{|z|=\varepsilon} \frac{d\bar{z}}{|\bar{z}| \log |z|} .$$

Using this and an elementary interchange of limits argument gives

$$\int_0^r \left\{ \int_{\Delta(\rho)} dd^c \log (\log a|f|^2)^2 \right\} \frac{d\rho}{\rho} = \int_0^r \left\{ \int_{\partial\Delta(\rho)} d^c \log (\log a|f|^2)^2 \right\} \frac{d\rho}{\rho}$$

$$= \int_0^r \rho \frac{d}{d\rho} \left(\frac{1}{2\pi} \int_{\partial\Delta(\rho)} \log (\log a|f|^2)^2 d\theta \right) \frac{d\rho}{\rho}$$

$$= \frac{1}{2\pi} \int_{\partial\Delta(\rho)} \log (\log a|f|^2)^2 d\theta + C .$$

Q.E.D.

To apply the S.M.T., we assume that Λ satisfies the curvature conditions

$$\text{Ric } \Lambda = \eta + \psi$$

where η is a closed, positive C^∞ (1,1) form and

$$\begin{cases} \psi \geq 0 \\ \psi^n \geq \Lambda . \end{cases}$$

The motivation for writing $\text{Ric } \Lambda$ in this manner is given by Proposition 2.17. We shall prove the following *Basic Estimate*:

(2.30) $$\int_0^r \left\{ \int_{B[\rho]} \eta \wedge \omega^{n-1} \right\} \frac{d\rho}{\rho} + N(R, r) \leq N(D, r) \qquad //.$$

PROOF. Following the same calculation as in the proof of (1.2),

$$\rho^{2n-2} \int_{B[\rho]} \psi \wedge \omega^{n-1} = \rho^{2n-2} \int_{B[\rho]} d(\psi \wedge d^c \log \|z\|^2 \wedge (dd^c \log \|z\|^2)^{n-2})$$

$$= \rho^{2n-2} \int_{\partial B[\rho]} \psi \wedge d^c \log \|z\|^2 \wedge (dd^c \log \|z\|^2)^{n-2}$$

$$= \int_{\partial B[\rho]} \psi \wedge d^c \|z\|^2 \wedge (dd^c \|z\|^2)^{n-2}$$

$$= {}^\circ\!\!\int_{B[\rho]} \psi \wedge \phi^{n-1}$$

where ϕ is the standard Kähler form on $\mathbf{C}^n$. Writing

$$\psi = \frac{\sqrt{-1}}{2\pi} \left(\sum_{i,j} \psi_{ij} \, dz_i \wedge d\bar{z}_j \right) ,$$

the non-negative Hermitian matrix (ψ_{ij}) satisfies

$$\frac{1}{n} (\text{Trace} \, (\psi_{ij})) \geq (\text{Det} \, (\psi_{ij}))^{\frac{1}{n}} \geq \lambda^{\frac{1}{n}} ,$$

the second step being the assumption $\psi^n \geq \Lambda$. Combining these two steps gives

$$\int_0^r \left\{ \int_{B[\rho]} \psi \wedge \omega^{n-1} \right\} \frac{d\rho}{\rho} = \int_0^r \left\{ \int_{B[\rho]} \psi \wedge \phi^{n-1} \right\} \frac{d\rho}{\rho^{2n-1}}$$

$$= \int_0^r \left\{ \int_{B[\rho]} \text{Trace} \, (\psi_{ij}) \Phi \right\} \frac{d\rho}{\rho^{2n-1}}$$

$$\geq \int_0^r \left\{ \int_{B[\rho]} \lambda^{\frac{1}{n}} \Phi \right\} \frac{d\rho}{\rho^{2n-1}}$$

$$= \int_0^r \left[\int \left\{ \int_{\partial B[\rho]} \lambda^{\frac{1}{n}} \Sigma \right\} t^{2n-1} \, dt \right] \frac{d\rho}{\rho^{2n-1}} .$$

Using $\text{Ric } \Lambda = \eta + \psi$ and setting

$$S(r) = \int_0^r \left\{ \int_{B[\rho]} \lambda^{\frac{1}{n}} \Phi \right\} \frac{d\rho}{\rho} \ ,$$

(2.31) $$\int_0^r \left\{ \int_{B[\rho]} \text{Ric } \Lambda \wedge \omega^{n-1} \right\} \frac{d\rho}{\rho} \geq \int_0^r \left\{ \int_{B[\rho]} \eta \wedge \omega^{n-1} \right\} \frac{d\rho}{\rho} + S(r) \ .$$

On the other hand, concavity of the logarithm gives

$$\int_{\partial B[r]} \log \lambda \cdot \Sigma \leq n \log \left(\int_{\partial B[r]} \lambda^{\frac{1}{n}} \Sigma \right)$$

$$= n \log \left\{ \frac{1}{r^{4n-2}} \left[r^{2n-1} \frac{d}{dr} \left(r^{2n-1} \frac{dS(r)}{dr} \right) \right] \right\}$$

$$\leq n\varepsilon \log r + n(1+\varepsilon) \log S(r) \qquad //_\varepsilon$$

by the calculus lemma 2.5. Since

$$S(r) \geq C \log r \quad (C > 0)$$

for large r, we find that

(2.32) $$S(r) - \int_{\partial B[r]} \log \lambda \cdot \Sigma \leq 0 \qquad //_\varepsilon \ .$$

Combining (2.31) and (2.32) with the S.M.T. (2.29) gives the basic estimate (2.30).   Q.E.D.

Bibliographical remarks

The importance of negative Gaussian curvature in value distribution theory seems to have been first recognized by F. Nevanlinna (cf. the references cited in [32]). Extensive use of differential-geometric methods was made by Ahlfors in [7] and [2]. It was he who first realized that the curvature condition $K \leq -1$ instead of $K \equiv -1$ was all that was necessary. The ubiquitous Ahlfors lemma appears in [2] for $n = 1$, with the extension to volume forms being due to Chern [12] and Kobayashi [28]. In a certain sense, this lemma founded the whole subject of *hyperbolic complex analysis*; cf. Kobayashi [29]. It also has extensive applications to algebraic geometry, especially to variation of Hodge structure; cf. [23] and the references cited there.

The proposition 2.16 on volume forms appears in Carlson-Griffiths [9]. The existence of such negatively curved volume forms was suggested by algebro-geometric considerations given in the thesis of Carlson [8]. The proof of the Big Picard Theorem using Nevanlinna was proved in a more general setting in Griffiths-King [22]. The Second Main Theorem is given in [9]. A general discussion of the role of negative curvature in the study of holomorphic mappings appears in [21]. The use of these methods in the geometrically somewhat more subtle *non*-equidimensional theory of entire holomorphic curves in P^n is given in Cowen-Griffiths [16], where references to this subject together with some additional heuristic discussion of negative curvature also appears. Regarding Picard-type theorems for holomorphic curves in general algebraic varieties, the recent compendium of examples by Green [19] is quite interesting. It was he who, together with Fujimoto, first proved the sharp Picard-type theorems for non-degenerate maps from C^m to P^n -- {hyperplanes} -- cf. the references given in [19].

CHAPTER 3

THE DEFECT RELATIONS

(a) *Proof of the defect relations*

Let $L \to M$ be a positive line bundle over a compact, complex manifold M. We consider the projective space $|L|$ of divisors of sections $s \in \mathcal{O}(M, L)$. An entire holomorphic mapping $f : C^n \to M$ is said to be *algebraically non-degenerate* in case the image does not lie in any $D \in |L|$. In this case the F.M.T. (1.15)

$$N_f(D, r) + m_f(D, r) = T_f(L, r) + C$$

and subsequent Nevanlinna inequality (1.16)

$$N_f(D, r) \leq T_f(L, r) + C$$

are valid. We may then define the *defect*

(3.1) $$\delta_f(D) = 1 - \varlimsup_{r \to \infty} \left[\frac{N_f(D, r)}{T_f(L, r)} \right] = \varliminf_{r \to \infty} \left[\frac{m_f(D, r)}{T_f(L, r)} \right]$$

with the properties:

(3.2) $$\begin{cases} 0 \leq \delta_f(D) \leq 1 \\ \delta_f(D_1) + \cdots + \delta_f(D_k) \leq k\delta_f(D_1 + \cdots + D_k) \\ \delta_f(D) = 1 \quad \text{if} \quad f(C^n) \quad \text{misses} \quad D \, . \end{cases}$$

$\Big($*Proofs*: The first follows from

$$0 \leq m/T \leq 1 + o(1) \, ,$$

65

and the third from $N_f \equiv 0$ in case $f(C^n) \cap D = \emptyset$. As for the second inequality, from Chapter 1

$$\frac{m_f(D_1, r)}{T_f(L, r)} + \cdots + \frac{m_f(D_k, r)}{T_f(L, r)} = k \frac{m_f(D, r)}{T_f(L^k, r)}$$

where $D = D_1 + \cdots + D_k \in |L^k|$. Now use $\Sigma(\underline{\lim}) \leq \underline{\lim}(\Sigma)$.$\Big)$

In general, *if the defect* $\delta_f(D)$ *is positive, then the image* $f(C^n)$ *meets* D *less often than is the case for an average divisor.* For example, suppose that M is projective space P^m and L is the hyperplane line bundle H. Then $|H|$ is the dual projective space P^{m*} of linear hyperplanes A, and Crofton's formula (1.7) together with "$\int \underline{\lim} \leq \underline{\lim} \int$" implies that

$$\int_{P^{m*}} \delta_f(A) \, dA = 0 \, .$$

To state the main result, we define the quantity

(3.3) $\delta(M, L) = \inf\{a \geq 0 : a c_1(L) + c_1(K_M) \geq 0\}$.

More precisely, we consider all $a \geq 0$ for which there are real closed (1,1) forms ϕ_L and ϕ_K representing $c_1(L)$ and $c_1(K_M)$ in $H^2_{DR}(M, R)$ such that

$$a \phi_L + \phi_K > 0 \, ,$$

and then $\delta(M, L)$ is the infimum of such a. For example, if $M = P^m$ and $L = H^d$ is the dth power of H, then

$$\delta(P^m, H^d) = \left(\frac{m+1}{d}\right) \, .$$

$\Big($*Proof:* We consider the unitary group $G = \mathcal{U}_{m+1}$ acting on P^m, and denote by dT the invariant measure on G. Given any differential form ψ on P^m, the average

$$\psi^* = \int_G T^*\psi \cdot dT$$

is well-defined, and gives an invariant form ψ^*. If ω is the standard Kähler form on $\mathbf{P}^m$, then any invariant form is a multiple $\lambda\omega^q$ of some power of ω. Thus, if

$$a\phi_H d + \phi_K > 0 \; ,$$

then by averaging

$$a\phi_H^* d + \phi_K^* > 0 \; .$$

But $\phi_H^* d = d\cdot\omega$ and $\phi_K^* = -(m+1)\omega$, so that $a > \frac{m+1}{d}$.   Q.E.D.$\Big)$

Given a non-degenerate, equidimensional holomorphic mapping $f: \mathbf{C}^n \to M$, we denote by R_f the ramification divisor of f and set

$$R_f(L) = \lim_{r\to\infty} \frac{N(R_f, r)}{T_f(L, r)} \; .$$

Then $R_f(L)$ measures the amount of ramification of f relative to its growth. The main result is

THEOREM (Defect Relation). *Suppose that* $f: \mathbf{C}^n \to M$ *is a non-degenerate, equidimensional holomorphic mapping,* $L \to M$ *is a positive holomorphic line bundle with*

$$c_1(L) + c_1(K_M) > 0 \; ,$$

and $D \in |L|$ *is a divisor with simple normal crossings. Then*

$$(3.4) \qquad\qquad \delta_f(D) + R_f(L) \leqq \delta(M, L) \; .$$

PROOF. Under the assumption $c_1(L) + c_1(K_M) > 0$ we may construct the volume form $\Psi_M(D)$ given by (2.16) and having the curvature properties as stated in Proposition 2.17. By the basic estimate (2.30), given $\varepsilon > 0$ we have an inequality

$$(1-\varepsilon)\,T_f(L\otimes K_M,r) + N(R_f,r) \leqq N(D_f,r) \qquad //_\varepsilon \ .$$

Rewriting this using the F.M.T. (1.15) and $T_f(L\otimes K_M,r) = T_f(L,r) - T_f(K_M^*,r)$ leads to

$$(3.5) \qquad \frac{m(D_f,r)}{T_f(L,r)} + \frac{N(R_f,r)}{T_f(L,r)} \leqq (1-\varepsilon)\,\frac{T_f(K_M^*,r)}{T_f(L,r)} + \varepsilon \qquad //_\varepsilon \ .$$

Noting that $\delta(M,L)\,c_1(L) \geqq c_1(K_M^*)$ implies

$$T_f(K_M^*,r) \leqq \delta(M,L)\,T_f(L,r) \ ,$$

we find

$$\varlimsup_{r\to\infty} \left[\frac{m(D_f,r)}{T_f(L,r)} + \frac{N(R_f,r)}{T_f(L,r)} \right] \leqq (1-\varepsilon)\delta(M,L) + \varepsilon \ .$$

Since $\Sigma\,(\varliminf) \leqq \varliminf(\Sigma)$ and $\varepsilon > 0$ was arbitrary, we have proved (3.4).

$$\text{Q.E.D.}$$

We now give some variants and special cases of (3.4). From (3.3) we note that

$$(3.6) \qquad \begin{cases} \delta(M,L) < \infty & \text{(since } L \to M \text{ is positive)} \\[2mm] \delta(M,L^N) = \dfrac{1}{N}\,\delta(M,L) & \text{(since } c_1(L^N) = Nc_1(L)) \ . \end{cases}$$

The condition

$$c_1(L) + c_1(K_M) > 0$$

is equivalent to

$$\delta(M,L) < 1 \ .$$

If, on the contrary $\delta(M,L) \geqq 1$, then trivially $\delta_f(D) \leqq 1 \leqq \delta(M,L)$. In summary: *Under the assumptions that* $L \to M$ *is positive,* $D \in |L|$ *has simple normal crossings, and* $f: C^n \to M$ *is non-degenerate and equidimensional*

$$(3.7) \qquad \delta_f(D) \leqq \delta(M,L) \ .$$

Here are some special cases:

(i) Assume that $D_1, \cdots, D_N \in |L|$ are divisors such that $D = D_1 + \cdots + D_N$ has simple normal crossings. Then

(3.8)
$$\sum_{i=1}^{N} \delta_f(D_i) \leq \delta(M, L) .$$

PROOF. Using the second property in (3.2), (3.6), and (3.7)

$$\sum_{i=1}^{N} \delta_f(D_i) \leq N\delta_f\left(\sum_{i=1}^{N} D_i\right)$$

$$\leq N\delta(M, L^N)$$

$$= \delta(M, L) .$$

(ii) Taking $M = P^n$ and $L = H^d$, we find that if $D_1, \cdots, D_N$ are smooth hypersurfaces of degree d meeting transversely, then

(3.9)
$$\sum_{i=1}^{N} \delta_f(D_i) \leq \left(\frac{n+1}{d}\right) .$$

(iii) If $d = 1$ in (3.9), we obtain

$$\sum_{i=1}^{N} \delta_f(A_i) \leq n+1$$

where $A_1, \cdots, A_N$ are hyperplanes in general position in P^n. For $n = 1$, the A_i are distinct points on the Riemann sphere, and (3.7) reduces to the beautiful *Nevanlinna defect relation*

(3.10)
$$\sum_i \delta_f(A_i) \leq 2 ,$$

which began this whole business.

(b) *The lemma on the logarithmic derivative*

In the preceding section we have used what might be termed the "method of negative curvature bounded away from zero on $M - D$" to obtain a lower bound on the size of $D_f = f^{-1}(D)$ for non-degenerate, equidimensional holomorphic mappings $f : C^n \to M$. Now, in R. Nevanlinna's original proof of the defect relation (3.10) for an entire meromorphic function $f(z)$, the crucial analytical step was the:

LEMMA ON THE LOGARITHMIC DERIVATIVE:

$$(3.11) \qquad \frac{1}{2\pi} \int_0^{2\pi} \log^+ \left| \frac{f'(re^{i\theta})}{f(re^{i\theta})} \right| d\theta \leqq C \left(\log T_f(r) + \log r \right) \quad // \ .$$

It was F. Nevanlinna who pointed out that (3.11) could also be proved by a negative curvature argument, but where the curvature is allowed to tend toward zero as one approaches D. What we shall do here is show that the same effect may be achieved by putting a metric on $M - D$ which is strongly negatively curved in a neighborhood of D, but which may have "positive curvature" outside such a neighborhood.

We shall be interested in the situation where the dual canonical bundle $K_M^* \to M$ is positive and where Λ is a meromorphic n-form on M with no zeroes and whose polar divisor $D \in |K_M^*|$ has simple normal crossings. The prototype example is $M = P^n$ so that $K_M^* = H^{(n+1)}$, and Λ is the differential form given in affine coordinates $(w_1, \cdots, w_n)$ by the Cauchy kernel

$$\Lambda = \frac{dw_1}{w_1} \wedge \cdots \wedge \frac{dw_n}{w_n} \ .$$

If $f : C^n \to M$ is a non-degenerate, equidimensional holomorphic mapping, then the pullback

$$f^* \Lambda = \Lambda_f = \lambda_f(z) dz_1 \wedge \cdots \wedge dz_n$$

where $\lambda_f(z)$ is an entire meromorphic function whose zero divisor is the ramification divisor R_f of f and whose polar locus is $D_f = f^{-1}(D)$. In case $n = 1$, $\lambda_f(z) = \dfrac{f'(z)}{f(z)}$ is the logarithmic derivative of the meromorphic function $f(z)$. We shall prove the estimate

(3.12)
$$\int_{\partial B[r]} \log^+ |\lambda_f| \Sigma = 0\,(\log\, T_f(K_M^*, r) + \log\, r) \qquad // \ ,$$

which reduces to (3.11) in the case of an entire meromorphic function $w = f(z)$ and $\Lambda = \dfrac{dw}{w}$.

Step one in the proof. Let η be a closed positive $(1,1)$ form on M which represents $c_1(K_M^*)$. As usual, we write $D = D_1 + \cdots + D_N$ where the D_i are smooth divisors meeting transversely, and we choose metrics in the line bundles $[D_i] \to M$ and sections $s_i \in \mathcal{O}(M, [D_i])$ such that $|s_i| \leq 1$ and $\displaystyle\sum_{i=1}^{N} dd^c \log \frac{1}{|s_i|^2} = \eta$. Letting Ψ_M be a volume form on M with $\mathrm{Ric}\,\Psi_M = -\eta$, we consider the singular volume form

$$\Psi_M(D) = \frac{\Psi_M}{\displaystyle\prod_{i=1}^{N} |s_i|^2 \,(\log\, a|s_i|^2)^2} \ .$$

The proof of Proposition 2.17 applied to this singular volume form gives: Given $\varepsilon > 0$, there is $\beta > 0$ such that for $0 < a \leq \beta$ the conditions

(3.13)
$$\begin{cases} \mathrm{Ric}\,\Psi_M(D) = \psi - \varepsilon\eta \\[2mm] \psi > 0 \quad \text{and} \quad \psi^n \geq \Psi_M(D) \end{cases}$$

are valid.

From the definition of $\Psi_M(D)$, if we write

$$(f^* \Psi_M(D))(z) = h(z)\Phi(z) \ ,$$

then for some constant $C > 0$

$$(3.14) \qquad \frac{1}{C} h(z) \leqq \frac{|\lambda_f(z)|^2}{\displaystyle\prod_{i=1}^{N} (\log |s_i|^2)^2} \leqq C h(z) .$$

In particular, if

$$\nu_f(r) = \int_{\partial B[r]} \log^+ |\lambda_f| \Sigma$$

is the quantity we wish to estimate, then

$$(3.15) \qquad \int_{\partial B[r]} \log h\Sigma \leqq 2\nu_f(r) + C' .$$

Step two in the proof. We apply the S.M.T. (2.29) to the singular volume form $h(z)\Phi(z)$, ignore the ramification term, and use (3.15) to obtain

$$\int_0^r \left(\int_{B[\rho]} \psi \wedge \omega^{n-1} \right) \frac{d\rho}{\rho} \leqq N_f(D, r) + \varepsilon\, T_f(K_M^*, r) + 2\nu_f(r) + C .$$

Now, arguing as in the proof of the basic estimate (2.30) we have

$$\int_0^r \left(\int_{B[\rho]} h^{\frac{1}{n}} \Phi \right) \frac{d\rho}{\rho^{2n-1}} \leqq \int_0^r \left(\int_{B[\rho]} \psi \wedge \omega^{n-1} \right) \frac{d\rho}{\rho} .$$

Combining these inequalities with (3.14) and the F.M.T. (1.15) gives

$$\int_0^r \frac{d\rho}{\rho^{2n-1}} \int_0^\rho t^{2n-1}\, dt \int_{\partial B[t]} \left[\frac{|\lambda_f|^{2/n}}{\prod (\log |s_i|^2)^{2/n}} \right] \Sigma \leqq C(T_f(K_M^*, r) + \nu_f(r)).$$

Applying the calculus lemma twice to the left-hand side of this estimate gives

$$(3.16) \quad \int_{\partial B[t]} \left(\frac{|\lambda_f|^{2/n}}{\prod (\log |s_i|^2)^{2/n}} \right) \Sigma \leqq C(T_f(K_M^*, r)^{1+\varepsilon} + r^\delta + \nu_f(r)) \quad //_{\varepsilon, \delta} \;.$$

We now take $\alpha = |\lambda_f|^{2/n}$ and $\beta = \prod_{i=1}^{N} \frac{1}{(\log |s_i|^2)^{2/n}}$ in the inequality

$$e^{\log^+ \alpha + \log \beta} \leqq \alpha\beta + 1 \;,$$

valid for $\alpha \geqq 0$ and $0 \leqq \beta \leqq 1$, to obtain from (3.16)

$$\int_{\partial B[r]} e^{\frac{2}{n} \log^+ |\lambda_f| - \frac{1}{n} \sum_{i=1}^{N} \log \left(\log \frac{1}{|s_i|^2} \right)^2} \Sigma \leqq C(T_f(K_M^*, r)^{1+\varepsilon} + r^\delta + \nu_f(r)) //_\delta \;.$$

Applying concavity of the logarithm, together with

$$\int_{\partial B[r]} \log \left(\log \frac{1}{|s_i|^2} \right)^2 \Sigma \leqq \log \left(\int_{\partial B[r]} \log \frac{1}{|s_i|^2} \Sigma \right)$$

$$= \log (m_f(D_i, r))$$

$$\leqq \log (T_f(K_M^*, r)) \;,$$

to the integral estimate yields

$$\nu_f(r) \leqq C (\log T_f(K_M^*, r) + \log r + \log \nu_f(r)) \quad // \;.$$

The last term on the right-hand side may be absorbed in the left-hand side, thereby proving (3.12).

(c) *R. Nevanlinna's proof of the defect relation 3.10*

For historical reasons, and also because it is very pretty, we shall use the lemma on the logarithmic derivative 3.11 to derive the defect relation for a non-rational entire meromorphic function $f(z)$. In this section we let

$$m_f(a, r) = \frac{1}{2\pi} \int_0^\infty \log^+ \frac{1}{|f(re^{i\theta}) - a|} \, d\theta \qquad (a \neq \infty)$$

$$m_f(\infty, r) = \frac{1}{2\pi} \int_0^\infty \log^+ |f(re^{i\theta})| \, d\theta \, ,$$

and shall use the Nevanlinna characteristic function

$$T_f(r) = N_f(\infty, r) + m_f(\infty, r) \, .$$

By Jensen's theorem, $T_f(r)$ is essentially independent of a (this is the F.M.T. for $T_f(r)$). We denote by $\varepsilon(r)$ any positive function of r satisfying

$$\varepsilon(r_n) = o(T_f(r_n))$$

for a sequence of values $r_n \to \infty$. According to Lemma 3.11,

(3.17) $m_{f'/f}(\infty, r) = \varepsilon(r) \, ,$

and $\log r$ is also $\varepsilon(r)$ since $f(z)$ was assumed non-rational.

Given distinct points $a_1, \cdots, a_q$ on the sphere P^1, which may each be assumed finite, obtaining a lower bound on the counting function sum

$$\sum_{\mu=1}^q N_f(a_\mu, r)$$

is, by the F.M.T., equivalent to finding an upper bound on the proximity function sum

$$\sum_{\mu=1}^q m_f(a_\mu, r) \, .$$

If we define the auxiliary function

$$g(z) = \sum_{\mu=1}^q \frac{1}{f(z) - a_\mu} \, ,$$

(1.19) and (3.17) together with

$$\log^+ |a+\beta| \leq \log^+ |a| + \log^+ |\beta| + \log 2$$

implies that

(3.18) $$m_{f'g}(\infty, r) = \varepsilon(r) .$$

On the other hand, the two expressions

$$
\begin{cases}
\displaystyle\sum_\mu m_f(a_\mu, r) = \frac{1}{2\pi} \int \log^+ \frac{1}{|f - a_\mu|} \, d\theta \\
\\
\text{and} \\
\\
\displaystyle m_{f'g}(\infty, r) + m_{f'}(0, r) = \frac{1}{2\pi} \int \left(\log^+ \left| \sum_\mu \frac{f'}{f - a_\mu} \right| + \log^+ \frac{1}{|f'|} \right) d\theta
\end{cases}
$$

don't look too different. If this is so, then

$$\sum_{\mu=1}^q m_f(a_\mu, r) = m_{f'}(0, r)^{+\varepsilon(r)}$$

where the right-hand side is independent of the points $\{a_\mu\}$. This is the idea behind the argument.

More precisely, for fixed $\delta > 0$ the points on the circle $|z| = r$ where $|f(re^{i\theta}) - a_\mu| \geq \delta$ contribute only an $0(1)$ term to $m_f(a_\mu, r)$. Choosing δ small relative to $\min_{\mu \neq \nu} |a_\mu - a_\nu|$, at the points where $|f(re^{i\theta}) - a_\mu| < \delta$ the contributions to $m_f(a_\mu, r)$ and $m_g(\infty, r)$ are the same, up to an $0(1)$ term. Repeatedly using the inequality

$$\log^+ |a - \beta| \leq \log^+ |a| + \log^+ |\beta| + \log 2 ,$$

we thus find an estimate

$$\sum_{\mu=1}^q m_f(a_\mu, r) \leq m_g(\infty, r) + 0(1).$$

On the other hand, from

$$\log^+ |\alpha\beta| \leq \log^+ |\alpha| + \log^+ |\beta|$$

we obtain

$$m_g(\infty, r) \leq m_{f'g}(\infty, r) + m_{f'}(0, r) ,$$

which, when combined with (3.18) and (3.19), yields

(3.20)
$$\sum_{\mu=1}^{q} m_f(a_\mu, r) \leq m_{f'}(0, r) + \varepsilon(r) .$$

Now add $N_{f'}(0, r)$ to both sides to find

$$\sum_{\mu=1}^{q} m_f(a_\mu, r) + N_{f'}(0, r) \leq T_{f'}(r) + \varepsilon(r) .$$

Using the F.M.T. for $f'(z)$,

$$T_{f'}(r) = N_{f'}(\infty, r) + m_{f'}(\infty, r)$$

$$\leq N_{f'}(\infty, r) + m_f(\infty, r) + \varepsilon(r)$$

by (3.17) again. Adding $2N_f(\infty, r)$ to both sides we obtain

(3.21)
$$\left\{ \begin{array}{l} \sum_{\mu=1}^{q} m_f(a_\mu, r) + N_1(r) \leq 2T_f(r) + \varepsilon(r) \\[2ex] \text{where} \\[1ex] N_1(r) = N_{f'}(0, r) + (2N_f(\infty, r) - N_{f'}(\infty, r)) . \end{array} \right.$$

The point here is that the term $N_1(r)$ is non-negative and measures the ramification of f (the points in $f^{-1}(\infty)$ where f is ramified are counted in a funny way). Thus

$$\sum_{\mu=1}^{q} m_f(a_\mu, r) \leqq 2T_f(r) + \varepsilon(r) \, ,$$

and this implies the defect relation.

Note that the "2" appeared essentially because on the sphere the differential dw has a pole of order two at $w = \infty$, which is the same reason as in the algebro-differential-geometric proof.

Here is an example of an entire holomorphic function $f(z)$ having $p + 1$ deficient values $b_0, \cdots, b_{p-1}, b_p = \infty$ with corresponding defects given by

(3.22)
$$\begin{cases} \delta_f(b_\mu) = \dfrac{1}{p} & (\mu = 0, \cdots, p-1) \\ \delta_f(b_p) = \delta_f(\infty) = 1 \, . \end{cases}$$

For this example, the Nevanlinna defect relation 3.10 is sharp. The general question of finding an entire meromorphic function $f(z)$ with pre-scribed defects $\delta_f(b)$ subject only to $\sum_{b \epsilon P^1} \delta_f(b) \leqq 2$ is quite interesting, especially when there are restrictions on the growth of f, and is a problem which has seen considerable progress in recent years.

The function in which we are interested is defined by

$$f(z) = \int_0^z e^{-t^p} dt \, .$$

Setting $t = \sigma e^{i\phi}$,

$$|e^{-t^p}| = e^{-\sigma^p \cos p\phi} \, .$$

It follows that in the angular sector

$$\Delta_\nu = \left\{ z : \left| \arg z - \frac{\nu\pi}{p} \right| < \frac{\pi}{2p} - \varepsilon \right\}, \quad (\nu = 0, 1, \cdots, 2p-1)$$

the integrand tends uniformly to zero for ν even and to infinity for ν odd. We shall correspondingly find that $f(z)$ tends to

$$b_\mu = e^{\frac{2\pi\mu\sqrt{-1}}{p}} \int_0^\infty e^{-t^p} dt = e^{\frac{2\pi\mu\sqrt{-1}}{p}} \Gamma\left(1 + \frac{1}{p}\right)$$

in $\Delta_{2\mu}$ $(\mu = 0, \cdots, p-1)$, and to

$$b_p = \infty$$

in $\Delta_{2\mu+1}$ $(\mu = 0, \cdots, p-1)$.

Indeed, in $\Delta_{2\mu+1}$ a double integration by parts gives

$$f(z) = -\frac{e^{-z^p}}{pz^{p-1}} (1+\delta) \qquad (\delta \to 0 \text{ as } |z| \to \infty) .$$

Thus in $\Delta_{2\mu+1}$

$$\log \frac{1}{|f(z)|} = r^p \cos p\phi(1+\delta) \qquad (z = re^{i\phi}) ,$$

from which it follows that

$$m_f(\infty, r) = \frac{r^p}{\pi} (1+\delta) \qquad (\delta \to 0 \text{ as } r \to \infty) .$$

Since $N_f(\infty, r) \equiv 0$, we have for the characteristic function

$$T_f(r) \equiv \frac{r^p}{\pi} (1+\delta) .$$

In the sector Δ_0,

$$f(z) - b_0 = \int_0^z e^{-t^p} dt - \int_0^\infty e^{-t^p} dt$$

$$= -\int_z^\infty e^{-t^p} dt$$

$$= \frac{-e^{-z^p}}{pz^{p-1}} (1+\delta) ,$$

where the last step is again integration by parts. Thus

$$\log \frac{1}{|\delta(z) - b_0|} = r^p \cos p\phi(1+\delta) \ ,$$

which gives

$$m_f(b_0, r) = \frac{r^p}{p\pi} (1+\delta) \ .$$

A similar result holds for $b_1, \cdots, b_{p-1}$ in the sectors $\Delta_2, \cdots, \Delta_{2p-2}$.
This implies the result (3.22) on the defects of f.

(d) *Ahlfors' proof of the defect relation 3.10*

As mentioned previously, it may be useful to give the two "classical" proofs of (3.10). The Ahlfors proof, which is based on the tension between the simultaneous relations

$$N_f(a, r) \leqq T_f(r) + 0(1)$$

(3.23)

$$\int_{a \,\epsilon\, \mathbf{P}^1} N_f(a, r) \, da = T_f(r) \ ,$$

is the quickest. Here, $T_f(r)$ is the Ahlfors-Shimizu characteristic function and da is the standard invariant measure on $\mathbf{P}^1$, normalized so that the total area is one. As mentioned in the discussion following equation (1.17), integration of the inequality in (3.23) over the range $f(C)$ gives a proof of the Liouville theorem that $f(C)$ is dense in $\mathbf{P}^1$. To obtain finer Picard-type theorems, one might try and apply the same method, but with a density having singularities at the points $a_1, \cdots, a_q$ in which we are interested. This is the idea.

Denoting by w the coordinate on $\mathbf{P}^1 = C \cup \{\infty\}$, the standard metric is

$$\Psi_{\mathbf{P}^1}(w) = \frac{\sqrt{-1}}{2\pi} \frac{dw \wedge d\bar{w}}{(1+|w|^2)^2} \ .$$

Assuming that the a_μ are finite, we consider

$$u(a_\mu, w) = \frac{|w - a_\mu|}{\sqrt{1 + |w|^2}},$$

which the reader will recognize as the length of the section $s_\mu \in \mathcal{O}(P^1, H)$ which defines a_μ. The density

$$\Psi(w) = \frac{C \Psi_{P^1}(w)}{\displaystyle\prod_{\mu=1}^{q} u(a_\mu, w)^{2\lambda}}$$

has singularities at the $\{a_\mu\}$ and is integrable for $\lambda < 1$. We choose the constant C so that the total area is one. Writing the mapping $f : C \to P^1$ as $w(z)$ and setting

$$f^*(\Psi_{P^1}) = \frac{\sqrt{-1}}{2\pi} h(z)\, dz \wedge d\bar{z},$$

we now integrate the inequality in (3.23) with respect to $\Psi(a)$ and obtain

$$\int_0^r \left(\int_{\Delta(\rho)} \frac{\sqrt{-1}}{2\pi} \frac{h(z)\, dz \wedge d\bar{z}}{\displaystyle\prod_{\mu=1}^{q} u(a_\mu, w(z))^{2\lambda}} \right) \frac{d\rho}{\rho} \leqq T_f(r) + 0(1).$$

Applying the calculus lemma 2.5 twice just as in step two of the proof of (3.12) gives

$$\frac{1}{2\pi} \int_0^{2\pi} \frac{h(re^{i\theta})\, d\theta}{\displaystyle\prod_{\mu=1}^{q} u(a_\mu, w(re^{i\theta}))^{2\lambda}} \leqq C(T_f(r)^{1+\varepsilon} + r^\delta) \qquad /\!/_{\varepsilon, \delta}.$$

The integrand may be written as

$$e^{\log h + 2\lambda \Sigma_{\mu=1}^q \log(1/u(a_\mu, w))},$$

and concavity of the logarithm gives

$$(3.25) \quad \frac{1}{2\pi} \int_0^{2\pi} h(re^{i\theta}) \, d\theta + \lambda \left(\sum_{\mu=1}^{q} m_f(a_\mu, r) \right) \leqq C(\log T_f(r) + \log r) \quad // \, .$$

The Gaussian curvature of $\Psi_{\mathbf{P}^1}$ is constant $+2$, so that the S.M.T. (2.1) may be used in (3.25) to yield

$$(3.26) \quad \lambda \left\{ \sum_{\mu=1}^{q} m_f(a_\mu, r) \right\} + N(R_f, r) \leqq 2T_f(r) + \varepsilon(r) \, ,$$

where

$$\lim_{r \to \infty} \left[\frac{\varepsilon(r)}{T_f(r)} \right] = 0 \, .$$

Dividing (3.26) by the order function and ignoring the ramification gives

$$\sum_{\mu=1}^{q} \lim_{r \to \infty} \left[\frac{m_f(a_\mu, r)}{T_f(r)} \right] \leqq \frac{2}{\lambda} \, .$$

Letting $\lambda \to 1$ yields the defect relation (3.10). Q.E.D.

(e) *A refinement in the classical case*

The reader may have noticed that much of the analysis in the theory as developed above is essentially of a one-variable character, and is centered around the distributional equation

$$\Delta \log |f| = \{f = 0\}$$

for a holomorphic function f. As might be expected, there are considerable refinements of the Nevanlinna defect relation in the classical situation of an entire meromorphic function of finite order. Generally speaking, these refinements appear to be based on the following two principles:

(i) *Such a function has an infinite product representation given by the Hadamard factorization theorem* (1.27). For example, if f(z) is an entire holomorphic function of finite non-integral order ρ, then the equation

(3.27) $$f(z) = a$$

has an infinite number of solutions for every $a \neq \infty$. (*Proof*: The function $f(z) - a$ has the same order ρ, and by (1.27)

$$f(z) - a = z^k e^{p(z)} \prod_\mu E(z/a_\mu, q)$$

where $q = [\rho]$, $p(z)$ is a polynomial, and the a_μ are the roots of (3.27). If the product were finite, then $f(z) - a$ would evidently have integral order.)

(ii) *Entire functions of finite order whose zeros and poles are on the negative and positive real axes respectively are frequently extremal for problems in value distribution theory.* This principle seems to be originally due to Gol'berg and has recently been put in beautiful conceptual form by Baernstein (cf. the bibliographical remarks below). It is of a different character from what we have encountered thus far, in that the angular — as opposed to the radial — distribution of the roots is involved.

In order to illustrate the flavor of these results, we shall prove a result of Edrei and Fuchs concerning an entire meromorphic function of finite order ρ with $0 \leq \rho \leq 1$. To state their theorem, we let a, b be two distinct points on the Riemann sphere and set

$$A = 1 - \delta_f(a)$$
$$B = 1 - \delta_f(b) \,.$$

On general grounds, all we can say is that

(3.28) $$0 \leq A, B \leq 1 \,,$$

i.e. (A, B) lies in the unit square in the first quadrant.

THEOREM (Edrei-Fuchs): *In addition to* (3.28),

(3.29) $A^2 + B^2 - 2AB \cos \pi\rho \geq \sin^2 \pi\rho$.

Moreover, if either $A < \cos \pi\rho$ *or* $B < \cos \pi\rho$, *then* $B = 1$ *or* $A = 1$
respectively.

COROLLARY (Valiron). *If* $\rho = 0$, *then* $f(z)$ *can have at most one deficient value.*

Thus functions of order zero have a striking similarity to rational
functions. The sort of restrictions imposed on (A, B) by this theorem are
illustrated by the shaded region in the following figure

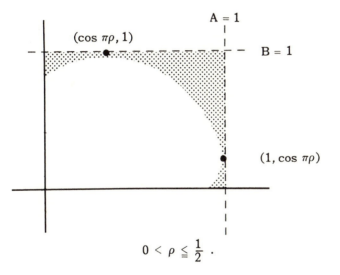

$$0 < \rho \leq \frac{1}{2} \ .$$

Step one in the proof: The result is trivial for $\rho = 1$, so we may assume
that $0 \leq \rho < 1$. Next, the function $g(z) = (f(z) - a)/(f(z) - b)$ has the same
order ρ as $f(z)$ and $\delta_g(0) = \delta_f(a)$, $\delta_g(\infty) = \delta_f(b)$. Consequently, we may
suppose that $a = 0$, $b = \infty$. Finally, multiplying $f(z)$ by Cz^k does not
change the problem, and allows us to assume that $f(0) = 1$.

By the Hadamard theorem,

$$f(z) = \frac{\prod_n (1 - z/a_n)}{\prod_m (1 - z/b_m)}$$

where the a_n, b_m are the respective zeros and poles of $f(z)$. We associate to $f(z)$ the symmetrized function

$$\hat{f}(z) = \frac{\prod_n (1 + z/|a_n|)}{\prod_m (1 - z/|b_m|)} \ ,$$

obtained by moving the zeros and poles of $f(z)$ to the negative and positive real axes along circular arcs. The first step is to prove the result, due to Gol'berg, that $f(z)$ is extremal for the present problem in the sense that

$$(3.30) \qquad \begin{cases} \delta_f(0) \leqq \delta_{\hat{f}}(0) \\[2mm] \delta_f(\infty) \leqq \delta_{\hat{f}}(\infty) \ . \end{cases}$$

Then, with the obvious notations and using (1.26),

$$\hat{A} \leqq A, \quad \hat{B} \leqq B \ ,$$

and

$$\rho = \inf \left\{ \lambda : \sum_n \frac{1}{|a_n|^\lambda} < \infty, \ \sum_m \frac{1}{|b_m|^\lambda} < \infty \right\} = \hat{\rho} \ ,$$

so that it will suffice to prove (3.29) for $\hat{f}(z)$. The proof of (3.30) is based on the

(3.31) PROPOSITION. *Given* $f(z)$ *and* $\hat{f}(z)$ *as above, and* $\phi(t)$ *a convex function of* $\log t$ *for* $t > 0$, *then for* $r \neq |a_n|$, $|b_m|$

$$\frac{1}{2\pi} \int_0^{2\pi} \phi(|f(re^{i\theta})|) d\theta \leqq \frac{1}{2\pi} \int_0^{2\pi} \phi(|\hat{f}(re^{i\theta})|) d\theta \ .$$

Proof of (3.30) from (3.31): Taking $\phi(t) = \log^+ t$,

$$m_f(\infty, r) \leqq m_{\hat{f}}(\infty, r)$$

for $r \neq |a_n|, |b_m|$, and then for all r by continuity. Since obviously

$$N_f(\infty, r) = N_{\hat{f}}(\infty, r) ,$$

$$\delta_f(\infty) = \underline{\lim} \left[\frac{m_f(\infty, r)}{m_f(\infty, r) + N_f(\infty, r)} \right] \leqq \underline{\lim} \left[\frac{m_{\hat{f}}(\infty, r)}{m_{\hat{f}}(\infty, r) + N_{\hat{f}}(\infty, r)} \right] = \delta_{\hat{f}}(\infty) ,$$

and similarly for $\delta_f(0)$ and $\delta_{\hat{f}}(0)$. Q.E.D.

The proof of (3.31) is, in turn, based on the following technical

(3.32) **LEMMA:** *Suppose that $\sigma_\nu(\theta)$ are even, periodic with period 2π, monotone increasing for $0 \leq \theta \leq \pi$, and satisfy $0 < m \leqq \sigma_\nu(\theta) \leqq M$. Then*

$$\int_{-\pi}^{\pi} \phi \left(\prod_{\nu=1}^{n} \sigma_\nu(\theta + \theta_\nu) \right) d\theta \leqq \int_{-\pi}^{\pi} \phi \left(\prod_{\nu=1}^{n} \sigma_\nu(\theta) \right) d\theta .$$

Proof of (3.32): By assumption the $\sigma_\nu(\theta)$ are *decreasing* functions of $\cos \theta$. Since

$$\cos (\theta - k) - \cos (\theta + k) = 2 \sin \theta \sin k$$

implies that, for $0 \leq \theta < \pi$,

$$\begin{cases} \cos (\theta - k) \geq \cos (\theta + k) & 0 \leq k < \pi \\ \cos (\theta + k) \geq \cos (\theta - k) & \pi \leq k < 2\pi , \end{cases}$$

we infer the inequalities

$$(3.33) \quad \begin{cases} \sigma_\nu(\theta - k) \leqq \sigma_\nu(\theta + k) & 0 \leqq \theta < \pi , \quad 0 \leqq k < \pi \\ \sigma_\nu(\theta + k) \leqq \sigma_\nu(\theta - k) & 0 \leqq \theta < \pi , \quad \pi \leqq k < 2\pi . \end{cases}$$

We next need to derive a consequence of the convexity of $\phi(e^t)$. If $\eta(t)$ is any convex function of t, then

$$\eta(t+s) + \eta(t-s)$$

is an increasing function of $s \geq 0$. Indeed, if $\eta(t)$ is smooth, then

$$\frac{d}{ds}\left(\eta(t+s) + \eta(t-s)\right) = \eta'(t+s) - \eta'(t-s) \geq 0 .$$

The general case follows by writing $\eta(t)$ as the limit of smooth convex functions. From this we deduce that, for t_1, t_2, s_1, s_2 non-negative,

$$\eta(t_1+t_2+s_1-s_2) + \eta(t_1+t_2+s_2-s_1) \leq \eta(t_1+t_2+s_1+s_2) + \eta(t_1+t_2-s_1-s_2).$$

Taking $\eta(t) = \phi(t)$ and setting

$$e^{t_1-s_1} = a \leq A = e^{t_1+s_1}$$

$$e^{t_2-s_2} = b \leq B = e^{t_2+s_2}$$

we find the inequality

(3.34)
$$\begin{cases} \phi(Ab) + \phi(aB) \leq \phi(AB) + \phi(ab) \\ 0 < a \leq A, \quad 0 < b \leq B \end{cases}$$

To prove the lemma, we may suppose that $0 \leq \theta_\nu < 2\pi$ and some $\theta_\mu > 0$. Define k_ν by $k_\nu \equiv \theta_\nu - \frac{1}{2}\theta_\mu(2\pi)$, $0 \leq k_\nu < 2\pi$, and set

$$\Sigma_1(\theta) = \prod_{0 \leq k_\nu < \pi} \sigma_\nu(\theta + k_\nu)$$

$$\Sigma_2(\theta) = \prod_{\pi \leq k_\nu < 2\pi} \sigma_\nu(-\theta + k_\nu) .$$

By periodicity and $\sigma_\nu(\theta) = \sigma_\nu(-\theta)$,

$$\int_{-\pi}^{\pi} \phi\left(\prod_{\nu=1}^{n} \sigma_\nu(\theta+\theta_\nu)\right) d\theta = \int_{-\pi}^{\pi} \phi\left(\prod_{\nu=1}^{n} \sigma_\nu(\theta+k_\nu)\right) d\theta$$

$$= \int_{-\pi}^{\pi} \phi(\Sigma_1(\theta)\Sigma_2(-\theta)) d\theta$$

$$= \int_{0}^{\pi} \{\phi(\Sigma_1(\theta)\Sigma_2(-\theta)) + \phi(\Sigma_1(-\theta)\Sigma_2(\theta))\} d\theta$$

$$\leq \int_{0}^{\pi} \{\phi(\Sigma_1(\theta)\Sigma_2(\theta)) + \phi(\Sigma_1(-\theta)\Sigma_2(-\theta))\} d\theta$$

$$= \int_{-\pi}^{\pi} \phi(\Sigma_1(\theta)\Sigma_2(\theta)) d\theta$$

where the next to the last step follows from (3.33) and (3.34). Combining gives

$$\int_{-\pi}^{\pi} \phi\left(\prod_{\nu=1}^{n} \sigma_\nu(\theta+\theta_\nu)\right) d\theta \leq \int_{-\pi}^{\pi} \phi\left(\prod_{\nu=1}^{n} \sigma_\nu(\theta+\theta'_\nu)\right) d\theta$$

where

$$\begin{cases} \theta'_\nu \equiv k_\nu - \frac{1}{2}\theta_\mu(2\pi), & 0 \leq \theta'_\nu < 2\pi, \quad \text{if} \quad 0 \leq k_\nu < \pi \\ \theta'_\nu \equiv -k_\nu - \frac{1}{2}\theta_\mu(2\pi), & 0 \leq \theta'_\nu < 2\pi, \quad \text{if} \quad \pi \leq k_\nu < 2\pi. \end{cases}$$

If $\theta_\nu = 0$, then $k_\nu = 2\pi - \frac{1}{2}\theta_\mu > \pi$ so that $\theta'_\nu = 0$. Moreover, $k_\mu = \frac{1}{2}\theta_\mu < \pi$ so that $\theta'_\mu = 0$. Proceeding in this way, after at most n steps we reach a situation where all $\theta'_\nu = 0$. Q.E.D.

Proof of (3.31): We may assume that

$$f(z) = \frac{\displaystyle\prod_{n=1}^{N} (1 - z/a_n)}{\displaystyle\prod_{m=1}^{M} (1 - z/b_m)}$$

and deduce the case of infinite products by passing to the limit. Setting

$$a_n = |a_n| e^{i\alpha_n}, \quad b_m = |b_m| e^{i\beta_m}$$

$$\sigma_\nu(\theta) = \left| 1 - \frac{re^{i\theta}}{|a_\nu|} \right|, \quad \theta_\nu = -\alpha_\nu \quad (\nu = 1, \cdots, N)$$

$$\sigma_{N+\mu}(\theta) = \left| 1 + \frac{re^{i\theta}}{|b_\mu|} \right|^{-1}, \quad \theta_{N+\mu} = \pi - \beta_\mu \quad (\mu = 1, \cdots, M)$$

in (3.32) gives the proposition.

Step two: We may assume that

$$f(z) = \left\{ \frac{\displaystyle\prod_{n=1}^{\infty} (1 + z/a_n)}{\displaystyle\prod_{m=1}^{\infty} (1 - z/b_m)} \right\}, \quad a_n, b_m > 0 .$$

The theorem is trivial unless

$$\delta_f(0) > 0 , \quad \delta_f(\infty) > 0 .$$

Assuming this, we may suppose that for large r

(3.35)
$$\left\{ \begin{array}{l} m_f(\infty, r) = \dfrac{1}{2\pi} \displaystyle\int_0^{2\pi} \log^+ |f(re^{i\theta})| \, d\theta > 0 \\[4mm] m_f(0, r) = \dfrac{1}{2\pi} \displaystyle\int_0^{2\pi} \log^+ \dfrac{1}{|f(re^{i\theta})|} \, d\theta > 0 . \end{array} \right.$$

The point in using the symmetrized function $f(z)$ above is that one may get some handle on the proximity functions as follows: First note that $|f(re^{i\theta})|$ is an even function of θ which decreases as θ goes from 0 to π — indeed, this is true for each of the factors

$$\left|1 + \frac{re^{i\theta}}{a_n}\right| , \quad \left|1 - \frac{re^{i\theta}}{b_m}\right|^{-1} .$$

By (3.35)

$$\log|f(-r)| < 0 < \log|f(r)| ,$$

and thus there is a unique $\beta(r)$ with $0 < \beta(r) < \pi$ such that

$$\log|f(re^{i\beta(r)})| = 0 .$$

From this we deduce

(3.36)

$$m_f(\infty, r) = \frac{1}{\pi}\int_0^{\beta(r)} \log|f(re^{i\theta})| \, d\theta$$

$$m_f(0, r) = \frac{1}{\pi}\int_{\beta(r)}^{\pi} \log\frac{1}{|f(re^{i\theta})|} \, d\theta .$$

LEMMA. *The integral representation*

$$(3.37) \quad T_f(r) = \frac{1}{\pi}\int_0^\infty N_f(0,t)\,\frac{r\sin\beta \, dt}{t^2 + 2tr\,\cos\beta + r^2} + \frac{1}{\pi}\int_0^\infty N_f(\infty,t)\,\frac{r\sin\beta \, dt}{t^2 - 2tr\,\cos\beta + r^2}$$

is valid.

PROOF. For z not on the real axis,

$$\log f(z) = \sum_n \log(1 + z/a_n) + \sum_m \log(1 - z/b_m)$$

$$= \int_0^\infty \log(1 + z/t)\,dn_f(0,t) + \int_0^\infty \log(1 - z/t)\,dn_f(\infty,t) .$$

Since $\frac{d}{dt}[\log(1+z/t)] = \frac{-z}{t(t+z)}$, $\frac{d}{dt}\left(\frac{1}{t/z}\right) = \frac{-1}{(t+z)^2}$, and $n_f(0,t) = O(t^{\rho+\varepsilon})$

where $\rho+\varepsilon < 1$, we may integrate by parts twice to obtain

$$\int_0^\infty \log(1+z/t)\,dn_f(0,t) = z\int_0^\infty \frac{1}{t+z}\frac{n_f(0,t)\,dt}{t}$$

$$= z\int_0^\infty \frac{dN_f(0,t)}{t+z}$$

$$= z\int_0^\infty \frac{N_f(0,t)\,dt}{(t+z)^2} \quad .$$

Doing the same for the polar terms gives

$$(3.38) \qquad \log f(z) = z\int_0^\infty N_f(0,t)\,\frac{dt}{(t+z)^2} + z\int_0^\infty N_f(\infty,t)\,\frac{dt}{(t-z)^2} \quad .$$

We will prove the lemma by combining (3.36) and (3.38). Thus the proximity function

$$m_f(\infty,r) = \frac{1}{\pi}\lim_{\delta\to 0}\int_\delta^{\beta(r)} \log|f(re^{i\theta})|\,d\theta$$

$$= \frac{1}{\pi}\lim_{\delta\to 0}\left[\operatorname{Im}\int_0^\infty N_f(0,t)\,dt \int_{re^{i\delta}}^{re^{i\beta}} \frac{dz}{(t+z)^2}\right.$$

$$\left. + \operatorname{Im}\int_0^\infty N_f(\infty,t)\,dt \int_{re^{i\delta}}^{re^{i\beta}} \frac{dz}{(t-z)^2}\right] \quad .$$

We set

$$P(t,r,\gamma) = \frac{1}{\pi}\operatorname{Im}\left(\frac{1}{t+re^{i\gamma}}\right) = \frac{r\sin\gamma}{t^2+2tr\cos\gamma+r^2}$$

$$Q(t,r,\gamma) = \frac{1}{\pi}\operatorname{Im}\left(-\frac{1}{t-re^{i\gamma}}\right) = \frac{r\sin\gamma}{t^2-2tr\cos\gamma+r^2} = P(t,r,\pi-\gamma) ,$$

and deduce that

$$m_f(\infty,r) = \int_0^\infty N_f(0,t) P(t,r,\beta) dt + \int_0^\infty N_f(\infty,t) Q(t,r,\beta) dt$$

$$- \lim_{\delta \to 0} \int_0^\infty N_f(0,t) P(t,r,\delta) dt - \lim_{\delta \to 0} \int_0^\infty N_f(\infty,t) Q(t,r,\delta) dt \; .$$

The first limit is obviously zero. As for the second, $Q(t,r,\delta) \to 0$ as $\delta \to 0$ uniformly in t outside any fixed interval $[r-\varepsilon, \; r+\varepsilon]$. Also, $N_f(\infty,t)$ is approximately equal to $N_f(\infty,r)$ inside such an interval. Thus, as $\delta \to 0$,

$$\int_0^\infty N_f(\infty,t) Q(t,r,\delta) dt = N_f(\infty,r) \int_{r-\varepsilon}^{r+\varepsilon} Q(t,r,\delta) dt + 0(1)$$

$$= N_f(\infty,r) \int_0^\infty Q(t,r,\delta) dt + 0(1)$$

$$= N_f(\infty,r) + 0(1)$$

since $\int_0^\infty Q(t,r,\delta) dt = 1$ for $\delta > 0$ by a residue calculation. Combining we find

$$T_f(r) = m_f(\infty,r) + N_f(\infty,r) = \int_0^\infty \{N_f(0,t) P(t,r,\beta) + N_f(\infty,t) Q(t,r,\beta)\} dt$$

as was to be proved.

Step three: The idea is that, according to (3.37) in the form

$$(3.39) \quad \begin{cases} T_f(r) = \displaystyle\int_0^\infty N_f(0,t) P(t,r,\beta) dt + \int_0^\infty N_f(\infty,t) P(t,r,\pi-\beta) dt \\[4mm] P(t,r,\gamma) = \dfrac{r \sin \gamma}{t^2 + 2tr \cos \gamma + r^2} \; , \end{cases}$$

the assumptions

$$\delta_f(0) > 0, \quad \delta_f(\infty) > 0$$

imply that both $N_f(0, r)$, $N_f(\infty, r)$ grow at a slower rate than $T_f(r)$, and we might hope to use this in (3.39) to force some sort of inequality. In order to carry this out, it is pretty clear that we will need some method for dealing with the possibly irregular growth of $T_f(r)$ — cf. Borel's lemma on page 29 and the calculus lemma 2.5. This is provided by the following

LEMMA ON POLYA PEAKS. *Given* $a(t)$, $\beta(t)$ *continuous positive functions of* $t \geq t_0$ *with the assumptions*

$$\begin{cases} \beta(t) \text{ is non-decreasing} \\ \overline{\lim_{t \to \infty}} \, a(t) = +\infty \\ \lim_{t \to \infty} \frac{a(t)}{\beta(t)} = 0 \, , \end{cases}$$

then there exist $r_n \to \infty$ *such that*

(3.40)
$$\begin{cases} a(t) \leq a(r_n), & t_0 \leq t \leq r_n \\ \dfrac{a(t)}{\beta(t)} \leq \dfrac{a(r_n)}{\beta(r_n)}, & r_n \leq t \, . \end{cases}$$

(*Such* r_n *are called Polya peaks.*)

PROOF. Suppose $t_1 > t_0$ is given. Then there is $t_2 > t_1$ such that

$$a(t_2) = \sup_{t_0 \leq t \leq t_2} a(t) \, .$$

Next, there is $t_3 > t_2$ such that

$$\frac{a(t_3)}{\beta(t_3)} = \sup_{t \geq t_3} \frac{a(t)}{\beta(t)} \, .$$

Finally, there is r with $t_2 < r < t_3$ such that

$$a(r) = \sup_{t_2 \le t \le t_3} a(t) = \sup_{t_0 \le t \le t_3} a(t) .$$

This implies (3.4) for $r_n = r$ since $\beta(t)$ is non-decreasing.   Q.E.D.

To apply this lemma, we choose $\varepsilon > 0$ sufficiently small, set $\lambda = \rho + \varepsilon < 1$, and take

$$a(r) = \frac{T_f(r)}{r^{\lambda - 2\varepsilon}}$$

$$\beta(r) = r^{2\varepsilon} .$$

We may then find arbitrarily large r such that

(3.41)
$$\begin{cases} \dfrac{T_f(t)}{t^{\lambda - 2\varepsilon}} \le \dfrac{T_f(r)}{r^{\lambda - 2\varepsilon}} , & t_0 \le t \le r \\[4mm] \dfrac{T_f(t)}{t^{\lambda}} \le \dfrac{T_f(r)}{r^{\lambda}} , & r \le t . \end{cases}$$

Now to the proof of theorem. Choose

$$\begin{cases} 1 > A_0 > A = 1 - \delta_f(0) \\ 1 > B_0 > B = 1 - \delta_f(\infty) , \end{cases}$$

so that for $t \ge t_0$

$$N_f(0, t) < A_0 T_f(t)$$

$$N_f(\infty, t) < B_0 T_f(t) .$$

Plugging these inequalities into (4.39) and using (4.41) gives, for arbitrarily large Polya peaks r,

$$T_f(r) \le T_f(r) \Bigg\{ A_0 \int_0^r P(t,r,\beta) \Big(\frac{t}{r}\Big)^{\lambda - 2\varepsilon} dt + A_0 \int_r^\infty P(t,r,\beta) \Big(\frac{t}{r}\Big)^{\lambda} dt$$

$$+ B_0 \int_0^r P(t,r,\pi - \beta) \Big(\frac{t}{r}\Big)^{\lambda - 2\varepsilon} dt + B_0 \int_r^\infty P(t,r,\pi - \beta) \Big(\frac{t}{r}\Big)^{\lambda} dt \Bigg\} + \varepsilon(r),$$

where the remainder term

$$\varepsilon(r) \leq N_f(0, t_0) \int_0^{t_0} P(t, r, \beta)\, dt + N_f(\infty, t_0) \int_0^{t_0} P(t, r, \pi-\beta)\, dt \ .$$

Clearly $\varepsilon(r) = \frac{0(1)}{r}$, and by making A_0, B_0 slightly larger may be absorbed in the term $\{\cdots\}$ above.

To estimate the integrals, we set $t = rs$ to have $0 \leq s \leq 1$, and then

$$\int_0^r P(t,r,\gamma)\left(\frac{t}{r}\right)^{\lambda-2\varepsilon} dt + \int_r^\infty P(t,r,\gamma)\left(\frac{t}{r}\right)^\lambda dt$$

$$= \int_0^1 s^{\lambda-2\varepsilon} P(s,1,\gamma)\, ds + \int_1^\infty s^\lambda P(s,1,\gamma)\, ds$$

$$= \int_0^\infty s^\lambda P(s,1,\gamma)\, ds + \int_0^1 (s^{\lambda-2\varepsilon}-s^\lambda) P(s,1,\gamma)\, ds \ .$$

An elementary contour integration gives

$$\int_0^1 s^\lambda P(s,1,\gamma)\, ds = \frac{\sin(\gamma\lambda)}{\sin(\pi\lambda)} \qquad (0 < \lambda < 1) \ .$$

Moreover, the other integral is $0(\varepsilon)$ for $\lambda \geq \lambda_0 > 0$. We deduce that

$$1 \leq \max_{0 \leq \beta \leq \pi} \left[\frac{A_0 \sin(\beta\lambda) + B_0 \sin(\pi-\beta)\lambda}{\sin \pi\lambda} \right] + C\varepsilon \ .$$

Choose $\gamma = \gamma(a_0, B_0, \lambda)$ to maximize

$$A_0 \sin \gamma\lambda + B_0 \sin(\pi-\gamma)\lambda \ .$$

For this value of γ,

$$(1-C\varepsilon)\sin \pi\lambda \leq (A_0 - B_0 \cos \pi\lambda)\sin \gamma\lambda + B_0 \sin \pi\lambda\gamma\lambda$$

$$\leq [(A_0 - B_0 \cos \pi\lambda)^2 + (B_0 \sin \pi\lambda)^2 \| \sin^2\gamma\lambda + \cos^2\gamma\lambda]$$

$$= A_0^2 + B_0^2 - 2A_0 B_0 \cos \pi\lambda$$

using the Cauchy-Schwarz inequality. Letting $\varepsilon \to 0$ and then $A_0 \to A$, $B_0 \to B$ gives (3.29).

To prove the remainder of the theorem, we suppose that $A < \cos \pi \rho$. Then $A < \cos \pi(\rho + \varepsilon)$ for sufficiently small ε. Taking $\lambda = \rho + \varepsilon$ we choose B_0 with $B < B_0 < 1$ and $A < B_0 \cos \pi \lambda$, and then set $A_0 = B_0 \cos \pi \lambda$ in (3.42) to obtain

$$(1 - C\varepsilon) \leqq B_0 \ .$$

We let B_0 decrease until either $B_0 \cos \pi \lambda = A$ or $B_0 = B$. Thus either

$$\frac{(1 - C\varepsilon)}{\cos \pi \lambda} \leqq A$$

or

$$(1 - C\varepsilon) \leqq B \ .$$

Letting $\varepsilon \to 0$ so that $\lambda \to \rho$, the first cannot hold, so that $B = 1$ as desired.   Q.E.D.

Bibliographical remarks

The defect relations, viewed as a quantitative form of Picard's theorem for an entire meromorphic function, are given in R. Nevanlinna ([31] and [32]). Subsequent proofs were found by F. Nevanlinna [32] and Ahlfors [1]. A modern exposition of Ahlfors' proof in a differential-geometric setting appears in Chern [13].

The equidimensional defect relation (3.4) is taken from Carlson-Griffiths [9]. In spirit, the proof is similar to that of F. Nevanlinna only with "curvature $\leqq -1$" replacing "curvature $\equiv -1$" as suggested by the Ahlfors lemma. In [22], an extension to equidimensional maps from an arbitrary algebraic variety to M is presented. Defect relations when D has worse singularities than simple normal crossings have been found by Shiffman [35], which also contains a somewhat different derivation of the First and Second Main Theorems. The thesis of Drouhillet [18] contains extensions to $f : C^n \to M_n$ of R. Nevanlinna's theorem on ramified values

and unicity of holomorphic mappings. Defect relations for an entire holomorphic curve in $\mathbf{P}^n$ are given in [16], where further references and some
historical discussion also appears.

The lemma on the logarithmic derivative is proved using the Poisson-
Jensen-Nevanlinna formula in [31], and another differential-geometric proof
by F. Nevanlinna appears in [32]. A weaker equidimensional version of
(3.7) is given in [22]. The proof of the classical defect relation from the
lemma on the logarithmic derivative follows the argument in [32], from
which the example $f(z) = \int_0^z e^{-t^p} dt$ is also taken.

The refinements of Nevanlinna theory in the classical case are due to
very many people. The book [25] by Hayman contains an extensive biblio-
graphy up through 1964. Our proof of the Edrei-Fuchs theorem closely
follows his. In recent years, there has been substantial progress on many
long-standing problems. The papers by Weitsman [42] and Drasin [17] contain a beautiful refinement of the Nevanlinna defect relation for functions
of finite order and a solution to the inverse problem (finding $f(z)$ with
pre-assigned defects $\delta(a)$ subject only to $\sum \delta(a) \leq 2$) respectively.
Finally, the papers [3] and [4] of Baernstein give an entirely new interpretation of the Nevanlinna characteristic function $T_f(r)$, probably leading to
conceptual proofs for the extremal behavior of symmetrized functions
$\hat{f}(z)$ and ultimately a much deeper understanding of the proximity integrals

$$\int_\alpha^\beta \log^+ |f(re^{i\theta})|\, d\theta \qquad (\alpha < \beta).$$

BIBLIOGRAPHY

[1] Ahlfors, L., *Über eine Methode in der Theorie der meromorphen Funktionen*. Soc. Sci. Fennica, vol. VIII, no. 10, pp. 1-14.

[2] _____ , *An extension of Schwarz's lemma*. Trans. Amer. Math. Soc., 43 (1938), pp. 359-364.

[3] Baernstein, A., *Proof of Edrei's spread conjecture*. Proc. London Math. Soc., 26 (1973), pp. 418-434.

[4] _____ , *Integral means, univalent functions and circular symmetrization*, to appear in Acta Math.

[5] Bishop, E., *Conditions for the analyticity of certain sets*, Mich. Math. J., 11 (1964), pp. 289-304.

[6] Bott, R., and Chern, S. S., *Hermitian vector bundles and the equidistribution of the zeros of their holomorphic sections*. Acta Math., Vol. 114 (1966), pp. 71-112.

[7] Carlson, J., *A remark on the transcendental Bezout problem*, Value-Distribution Theory (Part A), Marcel Dekker, New York (1974), pp. 133-143.

[8] _____ , *Some degeneracy theorems for entire functions with values in an algebraic variety*. Trans. Amer. Math. Soc., 168 (1972), pp. 273-301.

[9] Carlson, J., and Griffiths, P., *A defect relation for equidimensional holomorphic mappings between algebraic varieties*. Ann. of Math., vol. 95 (1972), pp. 557-584.

[10] _____ , *The order functions for entire holomorphic mappings*. Value-Distribution Theory (Part A), Marcel-Dekker, New York, (1974), pp. 225-248.

[11] Chern, S. S., *Complex manifolds without potential theory*, van Nostrand, Princeton, N. J. (1967).

[12] _____ , *On holomorphic mappings of Hermitian manifolds of the same dimension*. Proc. Symp. Pure Math., vol. 11 (1958), pp. 157-170.

[13] _____ , *Complex analytic mappings of Riemann surfaces I*. Amer. Jour. Math., 82 (1960), pp. 323-337.

[14] Cornalba, M., and Griffiths, P., *Some transcendental aspects of alge-braic geometry*. Proc. A.M.S. Summer Institute on Algebraic Geometry (1974).

[15] Cornalba, M., and Shiffman, B., *A counterexample to the "Trans-cendental Bezout Problem."* Ann. of Math., vol. 96 (1972), pp. 402-406.

[16] Cowen, M., and Griffiths, P., *Holomorphic curves and metrics of negative curvature*, to appear in Jour. d'Analyse.

[17] Drasin, D., *A meromorphic function with assigned Nevanlinna deficiencies*. Bull. Amer. Math. Soc., 80 (1974), pp. 766-768.

[18] Drouhillet, J., *Ramification and unicity of equidimensional holomor-phic maps*. Thesis at Rice University, Houston, Texas (1974).

[19] Green, M., *Some examples and counterexamples in value distribution theory for several variables*, to appear in Composito Math.

[20] Griffiths, P., *On the Bezout problem for entire analytic sets*, appeared in latest Annals.

[21] _____, *Differential geometry and complex analysis*. Proc. Symp. Pure Math., vol. 27 (1974), pp. 127-148.

[22] Griffiths, P., and King, J., *Nevanlinna theory and holomorphic map-pings between algebraic varieties*. Acta Math., 130 (1973), pp. 145-220.

[23] Griffiths, P., and Schmid, W., *Variation of Hodge structure (a discus-sion of recent results and methods of proof)*, to appear in Proc. Tata Institute Conference on Discrete Groups and Moduli.

[24] Gunning, R., and Rossi, H., *Analytic functions of several complex variables*, Prentice-Hall, Englewood Cliffs (1965).

[25] Hayman, W., *Meromorphic Functions*. Oxford Math. Monographs, Oxford University Press (1964).

[26] Kujala, R., *Functions of finite λ-type in several complex variables*. Trans. Amer. Math. Soc., 161 (1971), pp. 327-358.

[27] Kneser, H., *Zur theorie der gebrochenen funktionen mehrerer veränderlichen*. Iber. deutsch Math. Ver., 48 (1938), pp. 1-38.

[28] Kobayashi, S., *Volume elements, holomorphic mappings and the Schwarz lemma*. Proc. Symp. Pure Math., vol. 11 (1968), pp. 253-260.

[29] _____, *Hyperbolic manifolds and holomorphic mappings*. Marcel Dekker, New York (1970).

[30] Lelong, P., *Fonctions entières (n variables) et fonctions plurisousharmoniques d'ordre fini dans* C^n, J. d'Analyse 12 (1964), pp. 365-407.

[31] Nevanlinna, R., *Le théorèm de Picard-Borel et la theorie des fonctions méromorphes.* Gauthier-Villars, Paris (1929).

[32] —————— , *Analytic functions.* Springer-Verlag, Berlin and New York (1970).

[33] Santalo, L., *An introduction to integral geometry.* Hermann, Paris (1953).

[34] Shiffman, B., *On the removal of singularities of analytic sets.* Mich. Math. J., 15 (1968), pp. 111-120.

[35] —————— , *Nevanlinna defect relations for singular divisors,* to appear in Inventiones.

[36] Skoda, H., *Sous-ensembles analytiques d'ordre fini ou infini dans* C^n, Bull. Soc. Math. de France, vol. 100 (1972), pp. 353-408.

[37] Stoll, W., *Value distribution of holomorphic maps into compact complex manifolds.* Lecture Notes in Math. no. 135, Springer-Verlag (1970).

[38] —————— , *The growth of the area of a transcendental analytic set (I and II).* Math. Ann. 156 (1964), pp. 47-98; and 156 (1964), pp. 144-170.

[39] —————— , *About entire and meromorphic functions of exponential type.* Proc. Symp. Pure Math., 11 (1968), pp. 392-430.

[40] —————— , *Value Distribution Theory: Part B (Deficit and Bezout estimates),* Marcel-Dekker, New York (1974).

[41] Stolzenberg, G., *Volumes, limits, and extensions of analytic varieties.* Lecture Notes in Math. no. 19, Springer-Verlag, New York (1966).

[42] Weitsman, A., *A theorem on Nevanlinna deficiencies.* Acta Math. 128 (1972), pp. 41-51.

[43] Wells, R., *Differential analysis on real and complex manifolds.* Prentice-Hall, Englewood Cliffs, New Jersey (1973).

[44] Weyl, H., and Weyl, J., *Meromorphic functions and analytic curves.* Princeton University Press, Princeton (1943).

[45] Wu, H., *The equidistribution theory of holomorphic curves.* Princeton University Press, Princeton (1970).

ANNALS OF MATHEMATICS STUDIES

Edited by Wu-chung Hsiang, John Milnor, and Elias M. Stein

A complete catalogue of Princeton mathematics and science books, with prices, is available upon request.

PRINCETON UNIVERSITY PRESS

PRINCETON, NEW JERSEY 08540

Library of Congress Cataloging in Publication Data

Griffiths, Phillip, 1938-
    Entire holomorphic mappings in one and several com-
plex variables.

    (Annals of mathematics studies ; no. 85)
    1. Holomorphic mappings. I. Title. II. Series.
QA331.G753   515'.9   75-31631
ISBN 0-691-08171-9
ISBN 0-691-08172-7 pbk.